I0043839

Zootechnical
Beekeeping Management

Pablo Montesinos Arraiz

Northern Bee Books

Zootechnical beekeeping management
© Pablo Montesinos Arraiz 2024

Published in the United Kingdom by
Northern Bee Books,
Scout Bottom Farm,
Mytholmroyd,
West Yorkshire HX7 5JS
Tel: 01422 882751
Fax: 01422 886157

www.northernbeebooks.co.uk

ISBN 978-1-914934-84-1

Design and artwork, DM Design and Print

All figures and photos are by Pablo Montesinos Arraiz

Zootechnical
Beekeeping Management

Pablo Montesinos Arraiz

Contents

Chapter one.   **1**
Zootechnical beekeeping management.   1
Principles of the zootechnical beekeeping management.   8
Patterns of the brood nest.   15

Chapter two.   **20**
Method to estimate the number of frames with eggs, open brood,
sealed brood and bees in the brood chamber at the time of the
inspection of a hive.   20

Chapter three.   **52**
Methods of zootechnical evaluation of the queens.
    a. Method to determine the phenotype of the queens   56
    b. Method to evaluate the reproductive performance of the queens.   69

Chapter four.   **74**
Model of analysis, evaluation and control of the honey bees for
production purposes.   74

Chapter five.   **93**
Beekeeping coding system.   93
Method of hives inspection.   104
Beekeeping registers system.   112

Chapter six.   **119**
System of beekeeping indicators.   119

Chapter seven.   **130**
Animal welfare in beekeeping   130

Chapter eight.   **134**
Traditional beekeeping.   134

Preface

The growing demand for products from the beekeeping: honey, pollen, royal jelly, propolis, and venom, their importance as pollinating agents of agricultural crops for human consumption, and the production of seeds and feed for animal nutrition, and also the contribution of the honey bees in the genetic variability of wild vegetation ; make it necessary for bee farms to be managed following guidelines, principles and criteria similar to those used in the other livestocks.

There is also the incentive that honey bees can be considered as the farm animals with the least impact on the ecological and water footprint, in relation to the demand for inputs needed for its breeding.

Introduction

Honey bees have been extensively studied in their biology, ecology, anatomy, morphology, physiology, behavior and general beekeeping, and there is a vast literature on each of these aspects. However, the issues concerning honey bees in their condition as farm animals have not been deal with rigorous zootechnical criteria.

This book presents methods, models and systems related to how to gather, transcribe and organize in a systematic, methodological and orderly way, the information that comes from the biological behavior of honey bee colonies, expressed as beekeeping parameters, for their analysis, evaluation and control, to apply the techniques and/or methods of the general beekeeping management in the breeding of bees for production purposes. So, beekeeping farms should be managed with a professional, technical and business sense, in order to adapt them to the zootechnical, production and reproduction requirements, that improve and guarantee the viability and profitability of the beekeeping business, taking into account the conceptualization of animal welfare.

All of these methods,models and systems have been tested and used in the the Beekeeping Station of the Universidad Lisandro Alvarado, UCLA, Barquisimeto, Venezuela.

Chapter one

ZOOTECHNICAL BEEKEEPING MANAGEMENT

As all animal farms, the breeding of honey bees for production purposes should be managed with professional, technical and business guidelines, in order to adapt them to the zootechnical, production and reproduction requirements, that increase and contribute to the best operation and profitability of the beekeeping business.

Beekeeping farms should be developed as a system made up of elements, well integrated and coupled, in order to achieve the economic objectives that justify investments in the productive unit (Figure 1), always taking into account the Beekeeping Animal Welfare.

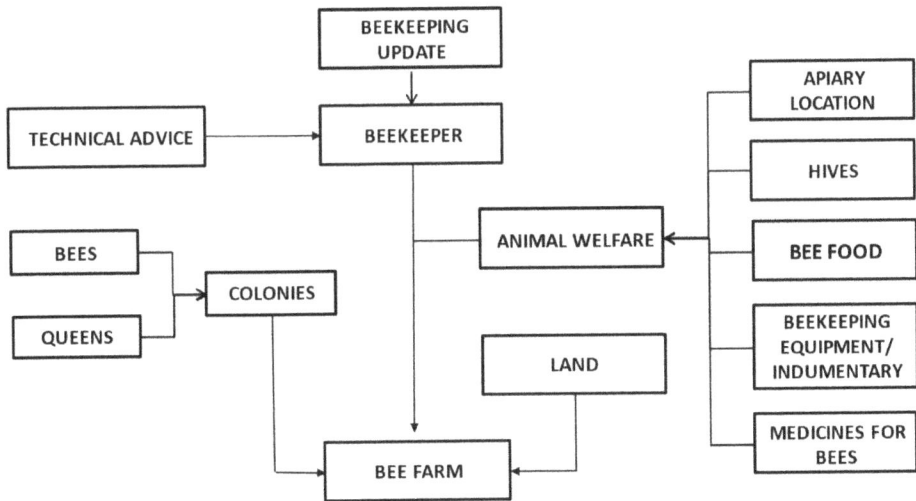

Figure 1. Beekeeping farm elements

The bees; main member of beekeeping farms, have the particularity that despite their status as insects (Figure 2), they are considered domestic and livestock animals, although great genetic, physiological and morphological changes have not been achieved in them as a consequence of their prolonged interaction with the man. However, some characters obtained through artificial selection and reproductive control in honey bees are heritable and constitute raw material for genetic improvement programs.

STATUS OF THE HONEY BEES IN THE ANIMAL KINGDOM
AND IN THE ZOOTECHNICS

```
        ┌─────────────────────────┐
        │      Invertebrate       │
        └─────────────────────────┘
                    │
                    ▼
        ┌─────────────────────────┐
        │        Arthropod        │
        └─────────────────────────┘
                    │
                    ▼
        ┌─────────────────────────┐
        │         Insect          │
        └─────────────────────────┘
                    │
                    ▼
        ┌─────────────────────────┐
        │  *Apis mellifera mellifera*  │
        └─────────────────────────┘
                    ▲
                    │
        ┌─────────────────────────┐
        │    Livestock animals    │
        └─────────────────────────┘
                    ▲
                    │
        ┌─────────────────────────┐
        │    Domestic animal      │
        └─────────────────────────┘
```

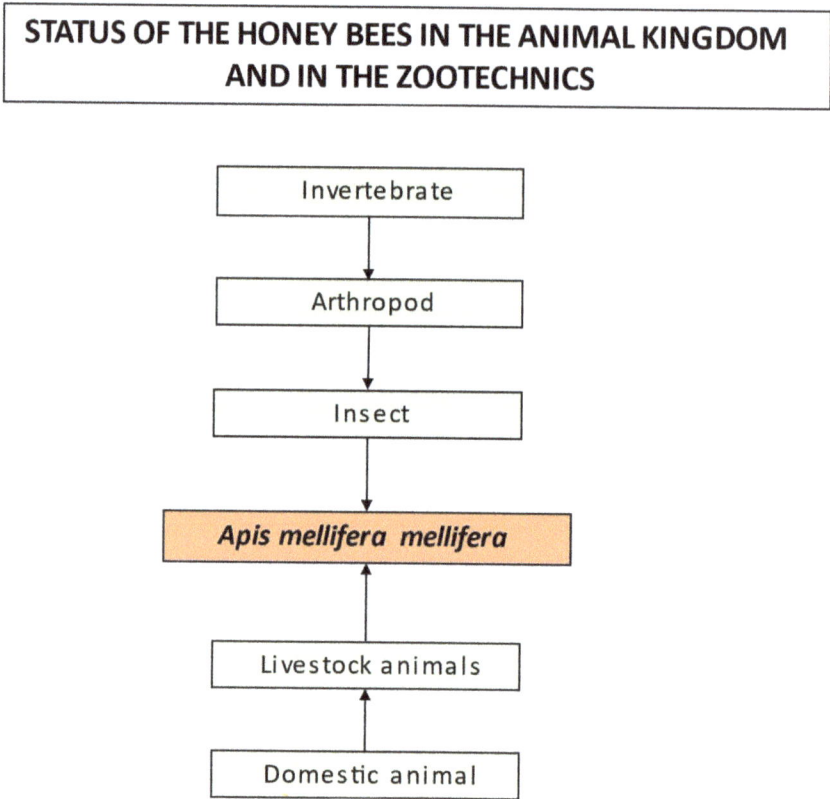

Figure 2. Status of the honey bees in the animal kingdom and in the zootechnics.

The management of the hives in small, medium, and large beekeeping farms is articulated around the time of year, population development, queen fecundity, and the nutritional status of the colonies. There are management variations in tropical countries, derived from its more benign climatic conditions, and almost permanent flowering during the summer and even with certain floral resources in the rainy season.

The determinants factors of the production in beekeeping (Figure 3) are related to the expression of the genetic production of honey bee colonies, which depend on the beekeeping management, and the environment influences. The first has to do with the Zootechnical beekeeping management and the General beekeeping management; and the second includes: climate, relief, deforestation, forest fires and biotic factors (living organisms). To them are added abiotic factors: water, temperature, light, air, pH and the soil with its nutrients. And also, pollution and the irrational use of biocides.

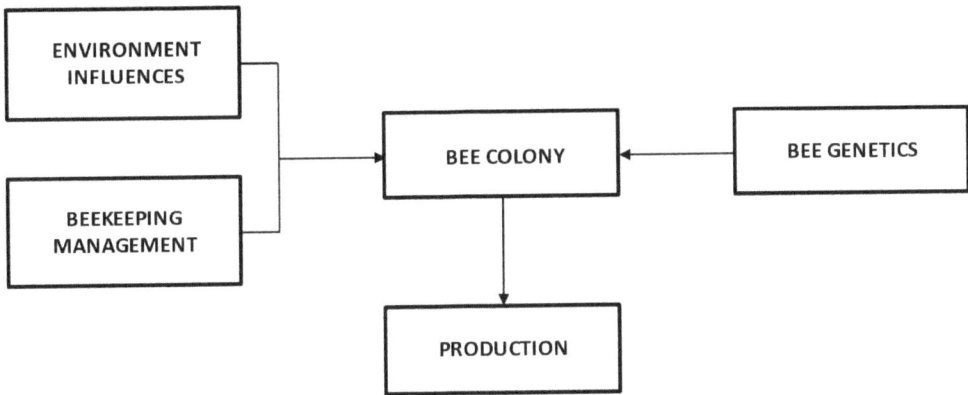

Figure 3. Production factors in beekeeping

The General beekeeping management has to with the methods and techniques to apply regarding the production, reproduction, health and nutrition issues, that the beekeepers carry out when working with the hives, according to the pre-harvest, harvest and post-harvest seasons, to obtain the maximum production of honey and by-products from the hives, always in correspondence with the animal welfare of bees.

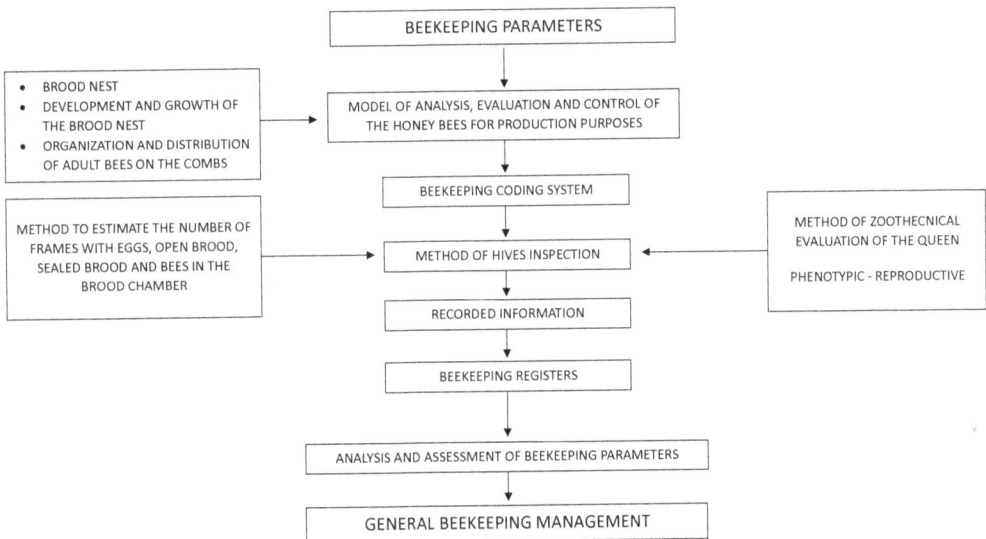

Figure 4. Zootechnical beekeeping management

The Zootechnical beekeeping management (Figure 4) refer to the analysis and evaluation of the beekeeping parameters that come from the principles related to the brood nest, the development and growth of the brood nest and the organization and distribution of adult bees on the combs (Figure 5), to apply the General beekeeping management, to obtain the maximum production of honey and by-products from the hives.

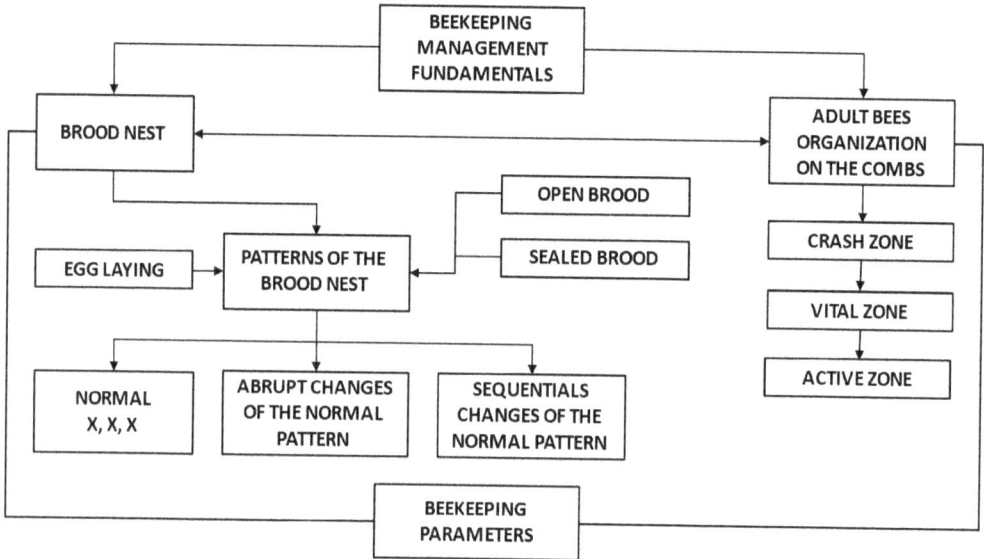

```
                        ┌─────────────────┐
                        │   BEEKEEPING    │
                        │   MANAGEMENT    │
                        │  FUNDAMENTALS   │
                        └─────────────────┘

┌──────────────┐                              ┌──────────────────┐
│  BROOD NEST  │◄─────────────────────────────►│   ADULT BEES     │
└──────────────┘                              │  ORGANIZATION    │
                                              │  ON THE COMBS    │
                       ┌──────────────┐        └──────────────────┘
                       │  OPEN BROOD  │
                       └──────────────┘        ┌──────────────────┐
┌────────────┐  ┌──────────────┐  ┌─────────┐  │   CRASH ZONE     │
│ EGG LAYING │  │ PATTERNS OF  │  │ SEALED  │  └──────────────────┘
└────────────┘  │  THE BROOD   │  │ BROOD   │
                │    NEST      │  └─────────┘  ┌──────────────────┐
                └──────────────┘               │   VITAL ZONE     │
                                              └──────────────────┘
┌──────────┐  ┌──────────────┐  ┌──────────────┐
│  NORMAL  │  │ ABRUPT       │  │ SEQUENTIALS  │  ┌──────────────────┐
│  X, X, X │  │ CHANGES OF   │  │ CHANGES OF   │  │   ACTIVE ZONE    │
└──────────┘  │ THE NORMAL   │  │ THE NORMAL   │  └──────────────────┘
              │ PATTERN      │  │ PATTERN      │
              └──────────────┘  └──────────────┘

                        ┌─────────────────┐
                        │   BEEKEEPING    │
                        │   PARAMETERS    │
                        └─────────────────┘
```

Figure 5. Beekeeping management fundamental

These parameters, that can be cualitative and quantitative, or only cualitative, are obtained through the Method of hives Inspection, and displayed through the Model of analysis, evaluation and control of the honey bees for production purposes, using the Beekeeping coding system.

The Zootechnical beekeeping management should precede and have primacy over the General beekeeping management, so in beekeeping farms the maximum production and productivity of the hives be achieved with optimal levels of profitability.

The success of the Zootechnical beekeeping management depends on the solidity and veracity of the information obtained, analyzed and evaluated by beekeepers, during and after routine inspections of the hives. This information, which must be recorded in the beekeeping registers, must include all the biological events of a zootechnical, productive and reproductive scope that take place in the hives, as well as all the circumstances of beekeeping nature that come about when working with the hives. Therefore, the importance of periodically inspecting the hives to do an individual and collective assessment and implement the techniques and/or methods of the General beekeeping management, that best suit each case. It is also necessary to monitor the effectiveness of the zootechnical beekeeping management, and the general beekeeping management, and also the yields achieved on the farm, for which zootechnical, production and reproduction beekeeping indicators and indices, should be used. On the other hand, in order to apply an appropriate Zootechnical and a General beekeeping management, it is a *sine qua non* condition to use beekeeping registers and enumerate the hives.

In beekeeping there are three types of parameters (Figure 6): Beekeeping Zootechnical Parameters, Beekeeping Production Parameters and Beekeeping Reproduction Parameters. The Beekeeping Zootechnical Parameters are classified into two groups: Zootechnical Behavioral Parameters and General Zootechnical Parameters.

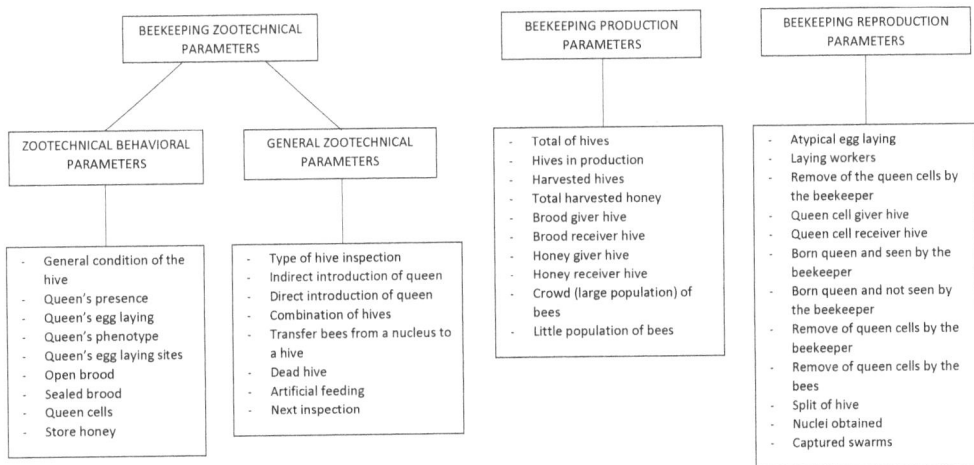

Figure 6. Types of beekeeping parameters.

The zootechnical behavioral parameters are those biological elements that are constant in the development and growth of honey bee colonies, and that are related to the development and growth of the breeding nest and the organization and distribution of the bees on the honeycombs.

There are nine zootechnical behavioral parameters:

1. General Condition of the hive (Good condition GC, Regular condition RC and Bad condition BC).
2. Queen's presence (Queen's presence QP, Queen's lacking QL and Orphan hive OH).
3. Queen's egg laying (abundant a, regular r, few f and null, and has egg laying HE).
4. Queen's phenotype (by color, by side and by age).
5. Queen's egg laying sites (in brood chamber E BC, in brood chamber and super E BCS and only in super E S).
6. Open brood OB (qualitative and quantitative).
7. Sealed brood SB (qualitative and quantitative).
8. Queen cells (open queen cell OQC and sealed queen cell SQC).
9. Honey stored (mature and immature, and quantitative and qualitative).

The general zootechnical parameters refer to the data or elements of the honey bees breeding that come from the general beekeeping management that is carried out in beekeeping operations.

There are eight general zootechnical parameters:

1. Type of hive inspection (Complete inspection CI and Incomplete inspection IC).
2. Indirect introduction of queen IIQ.
3. Direct introduction of queen DIQ.
4. Combination of hives A C B.
5. Transfer bees from a nucleus hive to a hive TNH.
6. Dead hive DH.
7. Artificial feeding AF.
8. Next inspection NI.

Regarding the beekeeping production and reproduction parameters, they have to do with the qualitative and quantitative or only qualitative variables, that result from the development and growth of honey bee colonies, and from which the beekeepers benefit in terms of products and by-products of the hives.

There are ten beekeeping production parameters:

1. Total of hives TH.
2. Hives in production HP.
3. Harvested hives HH
4. Total harvested honey THH
5. Brood giver hive BGH
6. Brood receiver hive BRH
7. Honey giver hive HGH
8. Honey receiver hive HRH
9. Crowd (large population) of bees CB
10. Little population of bees LB

There are twelve beekeeping reproduction parameters:

1. A typical egg laying AE
2. Laying workers LW
3. Remove of the queen by the beekeeper RQBB
4. Queen cell giver hive QCGH
5. Queen cell receiver hive QCRH
6. Born queen and seen by the beekeeper BQSB
7. Born queen and not seen by the beekeeper BQNSB
8. Remove of queen cells by the beekeeper RQCBB
9. Remove of queen cells by the bees RQCB
10. Split of hive SH
11. Nuclei obtained NO
12. Captured Swarms CS

PRINCIPLES OF THE ZOOTECHNICAL BEEKEEPING MANAGEMENT

In the Zootechnical beekeeping management, two fundamental principles should be considered: Relating to the queen and Relating to the workers. The first is based on the Sequential and orderly monitoring of the reproductive behavior of the queen, manifested through direct and indirect actions, and the second has to do with the organization and distribution of the workers on the combs (Figure 1).

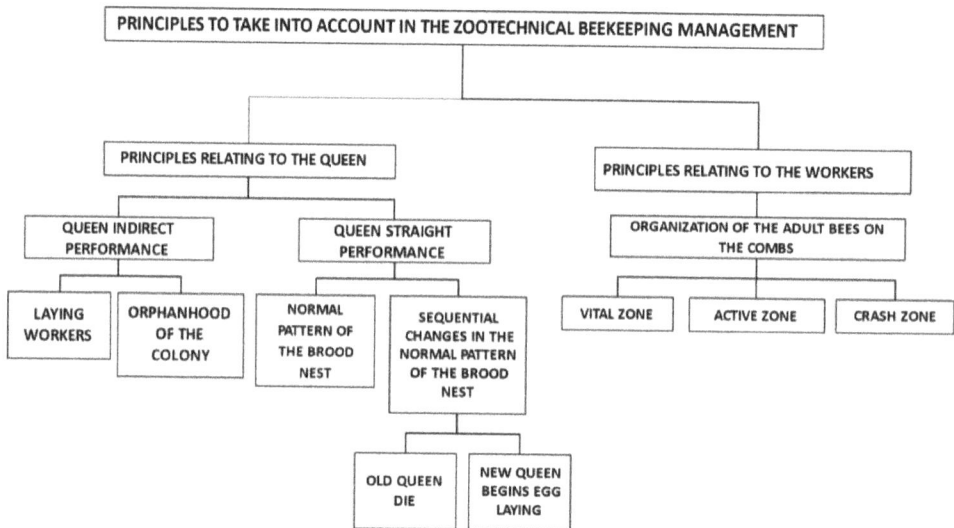

Figure 1. Principles of the zootechnical beekeeping management

▶ Principles relating to the queen
▶ Principles relating to the workers

PRINCIPLES RELATING TO THE QUEEN

The existence, development and perpetuation of the colony, depends on the first place on the queen; to which population growth and the storage of honey and pollen are closely linked. For this reason, it is necessary to look carefully at the phenotypic and reproductive behavior of the queen. The latter manifests itself directly and indirectly through the actions indicated below:

QUEEN STRAIGHT PERFORMANCE

- ▶ Normal brood nest pattern
- ▶ Sequential changes in the normal pattern of the brood nest:

  - • "Old" queen dies
  - • "New" queen starts egg laying

QUEEN INDIRECT PERFORMANCE

- ▶ Orphanhood of the colony
- ▶ Laying workers

QUEEN STRAIGHT PERFORMANCE

In the development of the brood nest there are a patterns that refer to the presence and/or absence of the elements that constitute the brood nest: eggs, open brood and sealed brood of the workers (see Patterns of the brood nest). Drones are reared in significant numbers only during colony breeding seasons.

QUEEN INDIRECT PERFORMANCE

Orphanhood of the colony

The death or disappearance of the queen is one of the most serious events that can happen in a colony. In the best of cases, orphanhood leads to a prolonged period without brood raising, therefore, there is no replacement generation for the dying bees. The population decreases as does the honey and pollen storage, and the colony becomes more susceptible to pillage and attack by predators.

To the extent that the normal workers decrease and the laying workers and consequently the drones increase, since these eggs, laid in worker and drone cells, being haploid will only give rise to drones; all the activities of the colony dwindle to nothing, the collection of nectar and pollen ceases, the food reserves are consumed. There are disorder and anarchy in the colony.

Consequently, the colony will die if the bees are unsuccessful in rearing a new queen, or the beekeeper does not introduce a new queen into the colony, or fails in the introduction technique.

The success of the workers in raising an emergency queen cell, when the original queen dies will depend on whether or not there is enough open brood and sealed brood, since there is a pheromone, glyceryl-1,2-diodelato-3-palmitate (Koeniger and Veith, 1983) which stimulates the rearing of queens in orphan colonies (Winston, 1991a), being the worker larvae less than 2 days old, the best to guarantee the viability and constitution of the new queens, instead of those larvae that are 4 to 5 days old. Sometimes, the workers may mistakenly choose drone larvae to raise queens, which will eventually be aborted (Winston, 1991b) or, if they are taken to sealed queen cells, the workers will not provide them with heat, and the larvae they contain die.

Laying workers

The queen and the open brood (Gary,2015; Winston, 1991a), and the sealed brood (Winston, 1991a), produce pheromones that inhibit the development of the worker ovaries. When the workers fail to raise a new queen and there is few or no brood, some workers become laying workers as their ovaries become slightly activated and they begin to lays eggs. This occurs 14 days after the colony is orphan, when there is no worker brood, and 30 days when the colony has brood (Winston, 1991b). In European races, laying workers appear between 23 and 30 days after the colony has lost its queen (Winston, 1991c). Once the laying workers appear in the colony, the normal workers become aggressive and fight among themselves, as well as making it difficult to raise queens and accept new queens (Sakagami, 1954). It should be mentioned, however, that laying workers can be found in normal colonies, especially in those with poorly queens (Kotepv, 1957) and in colonies showing signs of swarming (Tuenin, 1926) and Perepelova, 1928).

The laying workers have enlarged ovaries because some ovarioles have developed. Their eggs (actually ovules, because they have not been fertilized by the spermatozoa of the drones), smaller than those laid by the queens, are observed attached to the walls and rarely at the bottom of the cells, due to the short abdomen of the workers. The egg laying of the laying workers, on the other hand, is dispersed and disorderly (Winston, 1992b) Generally laying workers lay one egg per cell, although there can be up to six per cell from several laying workers (Winston, 1991b).

The behavior of the laying workers is very similar to that of the rest of the workers, in terms of the activities they carry out and as they are accepted; but as the drone brood spreads, the workers begin to show hostility towards them, pulling and pushing them (Sakagami, 1954; Hoffmann, 1961). Sometimes, a laying worker shows a slightly swollen and shiny abdomen, moving surrounded by workers who attend her, just as if it were a true queen (Sakagami, 1958; Velthuis et al, 1975; Crewe and Velthuis, 1980). This is probably due to an enlargement of its mandibular glands, which produce substances identical to those of a true queen (Velthuis et al, 1975; Crewe and Velthuis, 1980).

PRINCIPLES RELATING TO THE WORKERS

The activities of the workers are carried out within the framework of the following particularities (Gary, 2015; Winston, 1991d).

a)   In principle, the work of the workers obeys a temporal división according to age, governed in turn by the development and reabsorption of certain glands. The youngest workers, following a general progression of activities, will be found in the inner most parts of the bee nest, moving towards the outer most perimeters as they age, until finally leaving the nest to do field work.

b)   However, it must be pointed out that despite the fact that the physiological activities and the glandular development of the workers are genetically programmed and that is what defines the tasks at certain ages; It has been observed that certain conditions and requirements of the colony at a given moment, can induce changes in the ages at which the bees perform their activities, leading them to superimpose some tasks on others, except in the youngest bees (nurse) and in the older ones (foragers), in whom exclusive dedication to these tasks is observed. Also to prolong certain functions according to the needs of the colony in certain circumstances and even to perform several tasks at the same time. Occasionally there may be workers that carry out a single task during most of their lives.

c)   In the daily events of the colony, seven see, and perhaps surpassing the rest of their congeners, immobile bees on the honeycombs or, on the contrary, bees in a continuous walk throughout the nest. In the first case, beyond what could be thought of as lazy attitudes, they are actually producing larval food, wax or maturing honey; without prejudice to the fact that they may be resting at a certain moment. On the other hand, the uninterrupted movement of bees throughout the nest, perceived as meaningless and assigned, is intended to collect information about the requirements and needs that the colony may have, which will become the tasks to be implemented by the bees, recipients of such information.

d)   Adult bees are organized in the nest combs according to the functions performed and the particular circumstances of the colony at any given time. The bee population constitutes a dynamic living entity, which

moves changing its location and distribution on the combs, obeying the age they have and which govern the activities they perform.

The movement and location of the bees on the combs, appreciated as a unit composed of three layers, is an active and constant process, which depends on the climatic conditions, mainly the temperature and the greater or lesser in flow of nectar and pollen Which determines that the bee population opens or closes in the manner of an accordion, expanding or reducing its coverage in the brood nest, but always moving around the queen, to give her more or less space for egg laying.

Hence, adult bees can be observed on the honeycombs, in three zones in continuous harmonic movement, this is what is known as organization of adult bees on the combs (Sepulveda, 1980) Any alteration of the normality of the colony due to its members, breaks that gear, starting immediately in the colony, readjustment processes to return to its natural balance.

Thus, we can observe that the adult bees are organized on the honeycombs in three zones:

Vital zone

It is composed of the youngest bees (nurses). They surround the queen, to feed and groom her. They move to the adjacent spaces to previously clean the cells where the queen will move to perform her egg laying. The bees in this area are also dedicated to feeding and caring for the larvae and sealing the brood cells.

This is the area most sensitive to intrinsic and extrinsic changes that may affect the colony.

Active zone

It is made up of those workers dedicated to receiving the nectar and pollen brought by the foragers. They continue with the nectar digestion process before depositing it in the storage cells. They place the pollen in the respective cells, knead and compact it by adding digestive enzymes. They also build combs and remove the remains of caps from the cells from which newly hatched bees have emerged. They remove remains of wax, pollen, dead bees and other detritus from the hive to ensure the cleanliness of the hive.

Crash zone

This corresponds to the bees that occupy the most external and vulnerable combs of the nest. They are the oldest in the population and are committed to patrolling or touring the area to detect intruders who might have fooled the guards at the entrance. Also to receive information contained in the nectar and pollen brought by the foragers, and in response to it go out in search of those floral sources. Ventilation by the movement of their wings is another task of these workers. In this area are the bees that are starting their first orientation flights and then go out to forage. Finally, it is the assignment of the bees in the shock zone to be at the entrance of the nest as guardians and to go out to defend it against any threat that it may be subjected to.

References:

Koeniger, N., Veith, H.J., 1983. Glyceryl-1,2-dioleate-3-palmitate as a brood pheromone of the honey bee (*Apis mellifera* L.). Experientia 39:1051-1057

Winston, M.L., 1991a. The chemical world of honey bees. In, The biology of the honey bee. Ed Harvard University Press. Cambridge, Massachusetts, London, England, pp. 129-149.

Winston, M.L., 1991b. Development and Nutrition. In, The biology of the honey bee. Ed Harvard University Press. Cambridge, Massachusetts, London, England, pp. 46-71.

Gary, N.E., 2015. Activities and behavior of the honey bee. In, The hive and the honey bee. Editorial Dadant & Sons. Inc, Hamilton, IL, 62341, USA, pp. 247-337.

Winston, M.L., 1991c. Other worker activities. In: the biology of the honey bee. Ed Harvard University Press. Cambridge, Massachusetts, London, England, pp. 110-128

Sakagami, S.F., 1954. Occurrence of an aggressive behavior in queenless hive, with considerations of the social organization of honeybee. *Insectes Sociaux* 4:331-343.

Kotepv, V.S., 1957. Laying workers and swarming. Pchelovodstvo 34(6):31-32 (In Russian) (Apic. Abstr. 150/59).

Tuenin, T.A., 1926. Concerning laying workers. Bee World 8:90-91.

Perepelova, L.I., 1928. Laying workers, the egglaying- activity of the queen and swarming. Opytnaia Paseka 3:214-217 (In Russian); transl. In Bee World 10:69-71(1929).

Hoffmann, I., 1961. Ueber die Arbeitsteilungim weiselrichtigen und weisellosen Kleinvölkern der Honigbiene. Z. Bienenforsch. 5:267-279. (Apic. Abstr. 454/62).

Sakagami, S.F., 1958. The false queen: fourth adjustive response in dequeened honeybee colonies. *Behaviour.* 13:280-296.

Velthuis, H.H.W., Verheijen, F.J., Gottenbos, A.J., 1975. Laying worker honeybee: similarities to the queen. Nature. 207:1314.

Crewe, R.M., Velthuis, H.H.W., 1980. False queens: a consequence of mandibular gland signals in worker honeybees. Naturewissenschaften. 67:467-469.

Winston M.L., 1991d. The age-related activities of worker bees. In, The biology of the honey bee. Ed Harvard University Press. Cambridge, Massachusetts, London, England, pp. 89-109,

Sepulveda Gil, J.M., 1980. Enjambre, cría y reproducción. En, Apicultura. Editorial Aedos. Barcelona. España, pp. 55-70.

PATTERNS OF THE BROOD NEST

Basically, the Zootechnical beekeeping management and the General beekeeping management of the hives depend intrinsically on the development and growth of the brood nest (Figure 1) The brood nest has to do with the quantity, concentration, distribution and location of the eggs, the open brood and the sealed brood, that are mainly in the central combs of the bee nest (Figure 2). The lateral and upper combs will be occupied by reserves of pollen and nectar or honey. The brood nest also may extend into the lateral and upper combs of the bee nest, if there is not sufficient space for the queen to lay eggs and/or the amount of pollen and nectar collected by the bees is increasing.

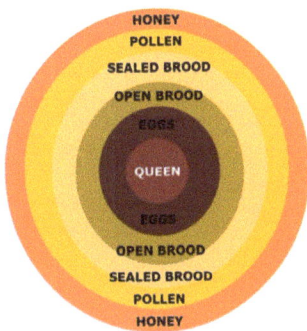

Figure 1. Brood nest

Figure 2. Bee nest

The patterns of the brood nest refer to the presence and/or absence of the elements that constitute the brood nest: eggs, open brood and sealed brood of the workers. Drones are reared in significant numbers only during breeding times of the colony. During the spring, summer, autumn and part of the winter, the colonies have in the brood nest the pattern X, X, X. The first X indicates the presence of eggs of one, two and three days, the second X of open brood (larvae in the different stages before being sealed); and the third X of sealed brood (larvae that have passed to the pupal stage). This pattern X, X, X in quantitative terms, increases from summer to autumn, until it reaches its maximum level, in relation to the abundance of nectar and pollen and the physical space that the colony occupies in the hive (Figure 3).

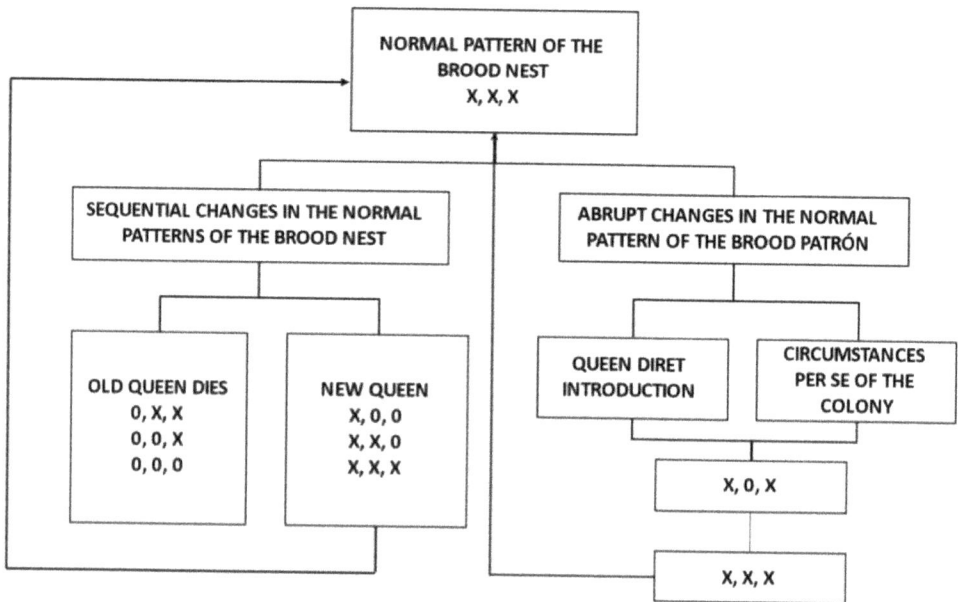

Figure 3. Paterns of the brood nest

NORMAL PATTERN OF THE BROOD NEST

X, X, X

In temperate region, at the end of autumn, eggs, open brood, sealed brood and the bee population begin to decrease because the queen is laying fewer and fewer eggs. The bee population is grouping around the queen to form the winter cluster. The bees bring less and less nectar and pollen to the hive. The queen in winter stops completely eggs laying, there will not be open brood and sealed brood either. It is observed the following sequential changes in the pattern of the brood nest:

O, X, X ↓

O, O, X ↓

O, O, O ↕

The normal pattern of the brood nest, X, X, X resumes as winter gives way to spring; this occurs gradually as follow:

X, O, O ↓
X, X, O ↓
X, X, X ↕

It is become present again, the cycle of quantitative development and growth of the brood nest and of the bee population. Honey and pollen will also progressively increase in the hive.

However, the queen can die unexpectedly and suddenly, due to a serious injury caused by the beekeeper at the time of inspectig the hive, by a predator, or by any disease. The death of the queen will bring with it, while the process of emergency queen rearing is carried out, changes in the sequence of the normal pattern of the brood nest.

The queen that has died is called "Old queen" because she is the one that the colony had in the last inspection of the hive; and the "New queen", comes from the emergency queen rearing. Once the new queen has reached sexual maturity and has mated, she will start laying eggs, with the consequent modifications in the patterns of the brood nest. Alterations in the patterns of the brood nest, triggered by queen rearing processes, swarming, supersedure; or queen replacement by the beekeeper are not considered here.

SEQUENTIAL CHANGES IN THE NORMAL PATTERN OF THE BROOD NEST,
FROM THE MOMENT THE OLD QUEEN DIES.

O, X, X ↓
O_2, X, X ↓
O_3, X, X ↓
O, O, X ↓
O, O, O ↕

The pattern represented by O, X, X, indicates total absence of eggs. The queen has died three days before. If there are two- and three-day-old eggs, O_2; it means that the queen has not been dead for more than 24 hours. The presence of only 3-day-old eggs, O_3; indicates the death of the queen 48 hours ago. To the first X, correspond those eggs that have already completed the incubation period (3 days) and have become larvae (open brood). If one -day-old larvae are found, it means that the queen has died four days ago; two-day-old larvae, five days ago; three-day-old larvae, six days ago; four-day-old larvae, seven days ago; and five-day-old larvae, the queen has died eight days ago. The larvae that have completed their development and have been sealed (sealed brood), are represented by the second X.

The pattern O, O, X, indicates the absence of eggs of one, two and three days, and also the absence of larvae (open brood); and   as more than eight days have elapsed and all larvae have been sealed, the X represents the sealed brood.

O, O, O is the pattern observed from the third week after the queen death. Beside no eggs and larvae there is not sealed brood because all pupae have transformed into adult bees, and have emerged from their cells. This pattern of the brood nest could continue if the colony is unsuccessful in raising a new queen.

When the old queen dies while the colony is in normal conditions, that is, it has open brood with larvae one to two days old, sealed brood, young bees (nurses) and honey and pollen reserves; the workers begin to rear a new queen in the first 5-6 hours (as is assumed here) after the death of the queen or from 12 to 48 hours later. To do this, the workers modify one or several worker cells that contain larvae less than 2 days old and which they feed with abundant royal jelly, transforming them into queen cups and then into emergency queen cells, from which the new queen will be born and that will be established after having killed other queens that could have been born.

SEQUENTIAL CHANGES IN THE NORMAL PATTERN OF THE BROOD NEST, FROM THE MOMENT THE NEW QUEEN BEGINS EGG LAYING.

Five days after birth, the queen reaches sexual maturity and leaves for her mating flights; generally mating on the first flight and laying eggs between the second and fourth day later. Here, it is assumed that the new queen   starts eggs laying three days after mating or eight days after being born, observing the pattern X_1, O, O, which indicates presence of eggs of one day, eggs of two days X_2, O, O and eggs of three days X_3, O, O and absence of open brood and of sealed brood. When the incubation period of the eggs ends (three days), and they become larvae, and are fed by the nurses (such feeding takes an average of 5.5 days); there will be in the brood nest the pattern X, X, O indicating eggs and open brood but absence of sealed brood. Eight days after the new queen has started egg laying, and   since the open brood has been sealed, the brood nest will have again, the pattern that had   before the death of the old queen: X, X, X eggs, open brood and sealed brood.

ABRUPT CHANGES IN THE NORMAL PATTERN OF THE BROOD NEST

If in an orphan hive like the one described above, a fertilized queen is introduced; as long as more than 8 but less than 21 days have elapsed in European bees, (duration of the development stages of the workers); the sequence of the aforementioned patterns will be modified. It will then be seen the pattern X, O, X, which shows the presence of egg and sealed brood, but not open brood; since the egg laying comes from the new queen introduced by the beekeeper and the sealed brood from the old queen that was in the hive. Later, the pattern X, O, X will change to X, X, X. So that, the eggs, the open brood and the sealed brood will belong to   the new queen introduced.

CIRCUMSTANCES *PER SE* OF THE COLONY

The X, O, X pattern can also be found in those cases in which the queen, for whatever reason (organic or functional disorders, lack of sufficient nurses or lack of food reserves), has stopped laying eggs for 8 days and the inspection of the hive is carried out just when the queen is reestablishing the egg laying and all the open brood has already been sealed.

Chapter two

METHOD TO ESTIMATE THE NUMBER OF FRAMES WITH EGGS, OPEN BROOD, SEALED BROOD AND BEES IN THE BROOD CHAMBER AT THE TIME OF THE INSPECTION OF A HIVE

When a beekeeper is inspecting the brood chamber of a hive, he wants to know how many frames are occupied by eggs, open brood, sealed brood and bees, to be aware of   the growth and development of the colony, and depending on the immediate, short-term, and future requirements that the hive may have, apply the appropriate methods and/or techniques of the General beekeeping management.

This method has to do with how to estimate the variables eggs, open brood, sealed brood and bees, relating them individually and numerically to the surfaces they occupy on both sides of each frame. (Langstroth hive with deep frame type).

Different kinds of methods have been used to determine the population of bees, brood and egg laying of the queen (Burgett & Burikam, 1985; Nasr et al; 1990; Cornelissen et al, 2009; Yoshiyama et al, 2011; Delaplane et al, 2013; Stan et al, 2021). These methods have to do with gravimetry, weighing the hives with and without bees; counting the frames with bees and brood; used of grids to count brood, and eggs; and photography and digital analysis. All of them are justified and have great value in their context; but they take time, disturb the activities of all members of the colony, damage the eggs, stress the bees and can damage or kill the queen. In addition, performing these tasks require special skills.

This is an easy, fast and noninvasive method of estimating the number of frames occupied by eggs, open brood, sealed brood and adult bees. It is based on the techniques used by Burgett & Burikam (1985) in the quantification of the number of adult workers on combs of standard size at full holding capacity, and the one used by Daubenmire (1959) in visual estimation of vegetative coverage and Rogers et al (1983) in sampling honeybee colonies for brood production. The possible effect of genetic and environmental factors on the said variables was not taken into consideration.

The side of the frame that is being inspected is mentally divided into four quadrants (Figure 1) that is, eight quadrants when considering the two sides of the frame. (Figure 2). Each side of the frame will have the variables eggs, open brood, sealed brood and bees or combinations of these variables. These variables are calculated independently of each other, by 25%, 50%, 75% or 100% (Figure 3, Figure 4), depending on the degree of occupation of each one of them in the quadrants of the side of the frame.

Figure 1. Quadrants of the side of a frame of a brood chamber

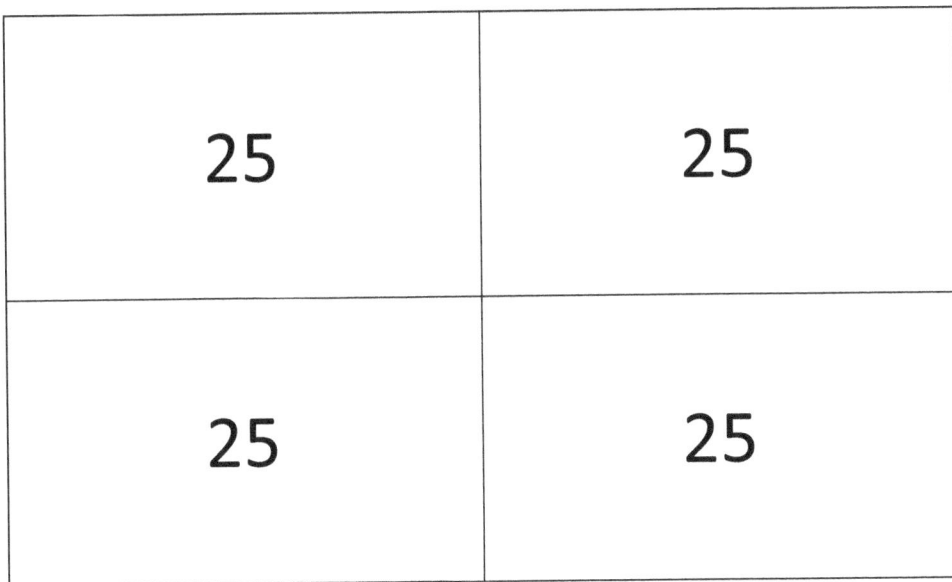

Figure 2. Quadrants of the side of a frame of the brood chamber that may be occupied by the variables eggs, open brood, sealed brood and bees.

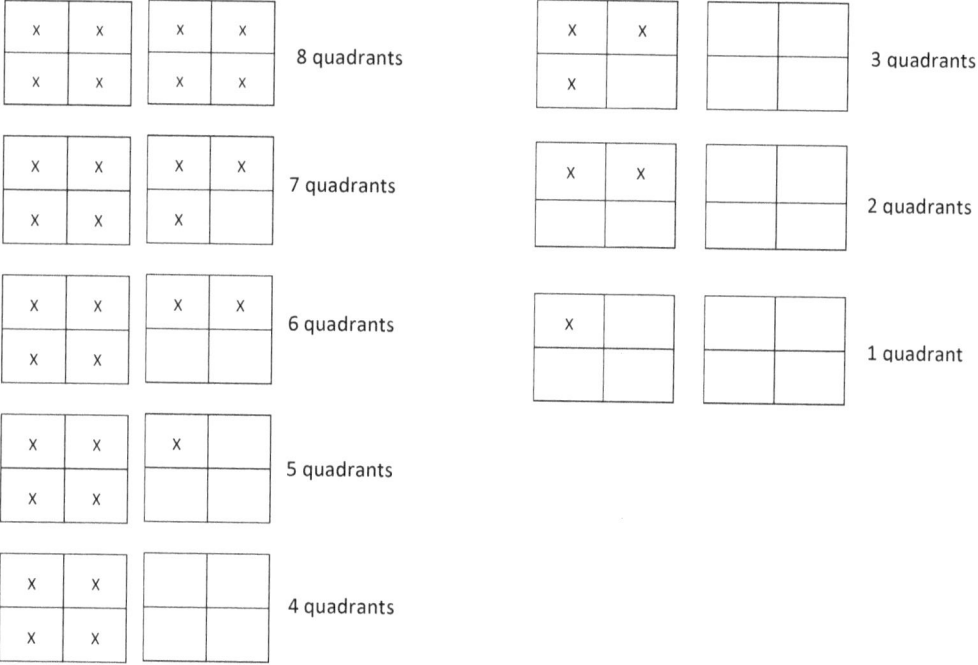

Figure 3. Possible occupation degrees in the frames of the brood chamber of the variables eggs, open brood and sealed brood

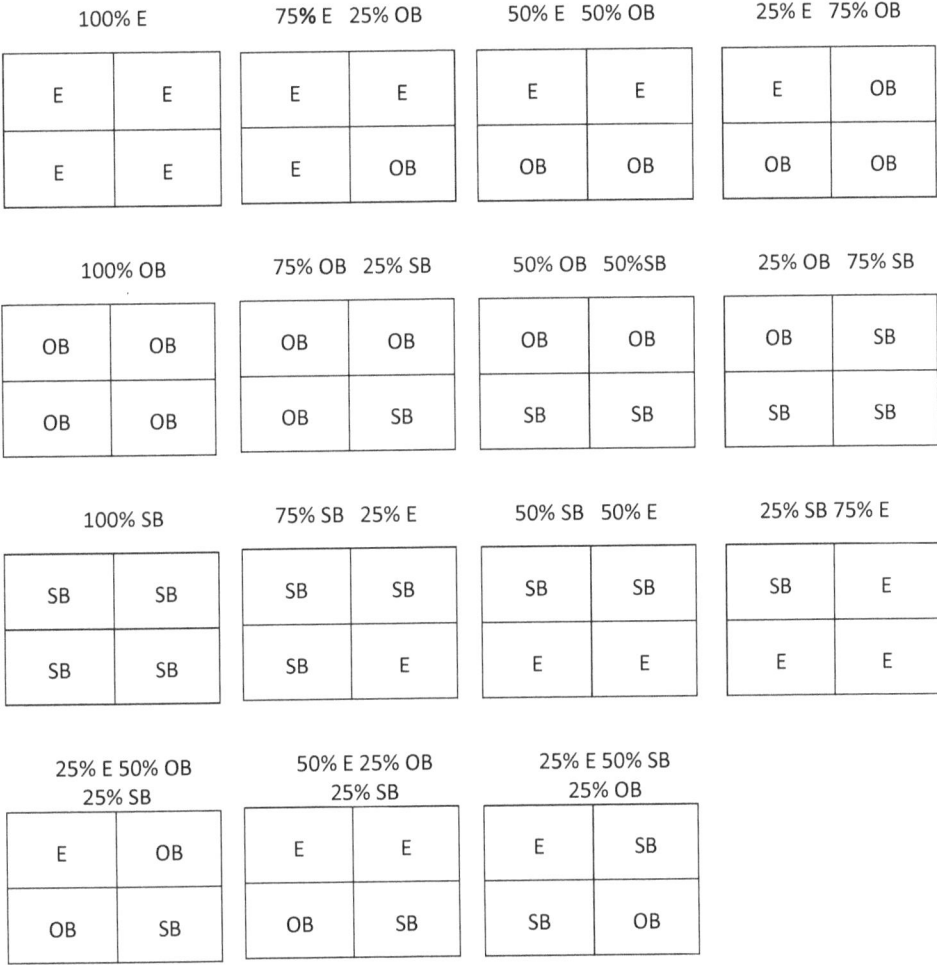

100% E	
E	E
E	E

75% E 25% OB	
E	E
E	OB

50% E 50% OB	
E	E
OB	OB

25% E 75% OB	
E	OB
OB	OB

100% OB	
OB	OB
OB	OB

75% OB 25% SB	
OB	OB
OB	SB

50% OB 50%SB	
OB	OB
SB	SB

25% OB 75% SB	
OB	SB
SB	SB

100% SB	
SB	SB
SB	SB

75% SB 25% E	
SB	SB
SB	E

50% SB 50% E	
SB	SB
E	E

25% SB 75% E	
SB	E
E	E

25% E 50% OB 25% SB	
E	OB
OB	SB

50% E 25% OB 25% SB	
E	E
OB	SB

25% E 50% SB 25% OB	
E	SB
SB	OB

Figure 4. Possible occupation degrees in the frames of the brood chamber of the variable bees in relation to the variables open brood, sealed brood and eggs

Bees %	Open brood %	Bees %	Sealed brood %	Bees %	Eggs %
25	25 50 75 100	25	25 50 75 100	25	25 50 75 100
50	25 50 75 100	50	25 50 75 100	50	25 50 75 100
75	25 50 75 100	75	25 50 75 100	75	25 50 75 100
100	25 50 75 100	100	25 50 75 100	100	25 50 75 100

When the variable being determined occupies the eight quadrants of the two sides of the frame, that is, 100%, the numerical equivalent in this case, one, is obtained by dividing the two sides of the frame by two (Table. 1).

Table 1. Quadrants of a frame, sides and or quadrants of a frame; frame; that can be covered by variable eggs, open brood, sealed brood or bees at the time of a hive inspection.

Quadrants of a frame	Sides and or quadrants of a frame	Divided by two	Frame
8	2 sides of a frame	1	1 a frame
7	1.75 one side of a frame and three quadrants of a frame	0.875	aprox. 1 a frame
6	1.50 one side of a frame and two quadrants of a frame	0.75	0.75 three quarters of a frame
5	1.25 one side of a frame and one quadrant of a frame	0.625	aprox. 0.75 three quarters of a frame
4	1.00 one side of a frame	0.50	0.50 half of a frame
3	0.75 three quadrants of a frame	0.375	aprox. 0.50 half of a frame
2	0.50 two quadrants of a frame	0.25	0.25 one quadrant of a frame
1	0.25 one quadrant of a frame	0.125	0.125 aprox. one quadrant of a frame

The decimal values are approximate to the values of the quadrants 25 -- 50 – 75 – 100

However, when eggs, open brood, sealed brood, or bees are scattered or concentrated in one or more quadrants on the side of the frame; such variables will be mentally regrouped by 25%, 50%, 75% or 100% (Figure 3, Figure 4).

Next, the side and / or the quadrants of the other side of the same frame that have the variable in question are added and dividing that quantity by two, gives us the numerical equivalent of the side and / or quadrants of that frame (Table 1). Then, what is obtained in the others frames of the brood chamber is added, obtaining a very approximate estimation of the number of frames with each of the variables.

One observation that could be made of this method would have to do with the inaccuracy of the beekeeper in regrouping the variables in the quadrants of the observed frames, but it has been tested at the same time with several beekeepers and the results tend to coincide with no significant differences (the beekeepers wrote down their estimates from the same hive at the same time, and then the results was compared). Another would be when the frames are overfull of bees but since the number of bees in the frames is not being estimated, but rather the number of frames covered by bees, this observation is not relevant in this method.

EXAMPLES OF ESTIMATION OF THE FRAMES WITH EGGS, OPEN BROOD, SEALED BROOD AND BEES IN THE BROOD CHAMBER AT THE TIME OF THE INSPECTION OF THE HIVE.

1) 50% Bees, 75% sealed brood

2) 100% Bees, 50% sealed brood

3) 25% Bees, 75% sealed brood

4) 25% Bees, 50% sealed brood

5) 25% Bees, 25% sealed brood

6) 75% Bees, 100% sealed brood

7) 100% Bees, 50% sealed brood

8) 100% Bees, 100% sealed brood

9) 25% Bees, 25% sealed brood

10) 50% Bees, 25% sealed brood

11) 25% Bees, 25% sealed brood

12) 25% Bees, 25% open brood, 25% sealed brood

13) 50% Bees, 50% sealed brood

14) 75% Bees, 25% sealed brood

15) 25% Bees

16) 75% Bees, 75% sealed brood

17) 75% Bees, 50% open brood

18) 100% Bees, 25% open brood, 25% sealed brood

19) 25% Bees, 25% sealed brood

20) 75% Bees

21) 75% Bees

22) 50% Bees

23) 25% Bees, 75% eggs, 25% sealed brood

24) 50% Bees, 50% sealed brood

25) 50% Bees, 50% sealed brood

26) 50% Bees, 75% sealed brood

27) 50% Bees, 75% sealed brood

28) 75% Bees, 75% sealed brood

29) 75 % Bees, 25% open brood, 25% sealed brood

30) 100 Bees, 25% open brood, 25% sealed brood

31) 25% Bees, 25% eggs, 25% open brood, 25% sealed brood

32) 25% Bees, 25% eggs, 75% sealed brood

33) 25% Beees, 25% eggs, 25% open brood, 25% sealed brood

34) 25% Bees, 25% eggs, 25% opend brood, 25% sealed brood

35) 100% Bees, 25% sealed brood

36) 50% Bees, 50% eggs, 25% opend brood, 25% sealed brood

37) 100% Bees, 25% open brood, 25% sealed brood

38) 25% Bees, 50% eggs, 25% open brood, 25% sealed brood

39) 50% Bees, 25% eggs; 50% open brood, 25% sealed brood

40) 25% Bees, 75% open brood, 25% sealed brood

41) 25% Bees, 25% open brood, 25% sealed brood

42) 100% Bees, 75% sealed brood

43) 50% Bees, 75% sealed brood

44) 100% Bees, 75% sealed brood

45) 100% Bees

46) 100% Bees, 25% eggs, 25% open brood, 25% sealed brood

47) 100% Bees, 50% sealed brood

48) 100% Bees

49) 100% Bees

50) 100% Bees

References

Burgett, M; Burikam, I. 1985. Number of Adult Honey Bees (Hymenoptera: Apidae) Occupying a comb: A Standard for Estimating Colony Populations. *J Econ Entomol.* 78: 1154-1156.

Cornelissen, B; Blacquière, T; Van Der Steen, J. 2009. Estimating honeybee colony size using digital photography. In Proceeding of 41st International Apicultural Congress, Montpellier, France. p. 48.

Daubenmire, H. 1959. A canopy-coverage method of vegetational analysis. *Northwest Sci.* 33: 43-64.

Delaplane, K. S; Van Der Steen, J; Guzman-Novoa, E. 2013.Standard methods for estimating strength parameters of *Apis mellifera* colonies. *Journal of Apicultural Research.* 52(1), 1-12.

Nasr, M. E; Thorp, R.W; Tyler, T.L; Briggs, D.L.1990. Estimating Honey Bee (Hymenoptera: Apidae) Colony Strength by a Simple Method: Measuring Cluster Size. *Journal of Economic Entomology.* 83(3), 748–754.

Rogers, L. E., R.O. Gilbert; Burgess. M. 1983. Sampling honeybee colonies for brood production: a double sampling technique. *J. Apic Res.* 22: 282- 241.

Stan, Ch; Requier, F; Chadoeuf, J; Guilbaud, L. 2021. Rapid measurement of the adult worker population size in honeybee. *Ecological Indicator.*122(4).

Yoshiyama, M; Kimura, K; Saitoh, K; Iwata, H. 2011. Measuring colony development in honey bees by simple digital analysis. *Journal of Apicultural Research.* 50(2): 170-172.

Chapter three

METHODS OF ZOOTECHNICAL EVALUATION OF THE QUEENS

In the genetic selection and improvement programs of the reproductive animals, an assessment must be made of the quantitative or measurable characteristics whose hereditary differences are transmitted from generation to generation, by the same mechanisms of the genes responsible for the qualitative differences (Shrode, 1980).

A reproductive animal can be evaluated through the analysis of several aspects: assessment of its genetic potential at the request of the qualities of its ancestors (pedigree), by that of its half-siblings (collaterals), by that of its descendants (testing), or through their own genetic material (studying their genome); and it can also be evaluated through its productive performance and its phenotype (Sañudo et al, 2009).

Despite the fact that honey bees were the first intensively breeding animals, after the silkworm *Bombyx mori* L (Borror et al, 1989) and that the first research carried out on intensive livestock farming was directed at beekeeping (Shrimpton, 1970), there have not set criteria, patterns and parameters to describe the queens bees phenotype and neither the reproductive performance, as has been done in other livestock farms with their breeding animals. All of which bring about a great void in the queens evaluation.

The races of bees have been differentiated based on biometric methods and behavioral characteristics (Ruttner, 2015). Biometric measurements have to do with the width of the thorax and abdominal segments; the length of the tongue, legs and wings; the color of the first segment dorsal abdomen; the length of the tongue; the hairy covering and the wings venation (Ruttner, 1975, Daly and Balling, 1978; Rinderer et al, 1986). Thus we see that the biometric methods have an eminently entomological approach and basically oriented to the bee workers, not giving importance to the zootechnical-reproductive characteristics of queens.

In the queens has been studied: the relationship of weight with the number of ovarioles (Hoopingarner and Farrar, 1959); how the fecundity of queens is influenced by their weight (Boch, 1960); the relationship among the weight of the queens at birth with the number of ovarioles and the volume of the spermatheca (Corbella, 1982); how the selection for the width of the abdomen of the queens improves some production characteristics (Costa, 2005); the correlation among the genetic and phenotype parameters and the weight, width and length of the abdomen (Halak, 2012).

From a zootechnical point of view, of queens in general it has only been said that those with large abdomen, rounded flanks that gradually thin out and have uniform color are good for laying eggs, although sometimes a queen with the characteristics described above is not necessarily a good egg-laying queen (Cale et al, 2015).

The criteria used to assess the behavioral characteristics of the queens are inferred from the behavior of the colonies (of their progeny) the ability to winter, the degree of docility, the tranquility on the combs when the hives are been inspecting, the non-willingness to swarm, or that if it is a good honey-producing colony queen (Cale et al, 2015).

There have been indicated of the queens imprecise aptitudes, such as: whether she has filled three or four combs with brood or whether the egg laying is concentric and concentrated and with brood of similar ages, she is a good queen; that if she is erratic in her movements it is not desirable; that if she lays continuously, producing brood through out the season and into late fall, she is a good queen (Cale et al, 2015).

Hence, the absence of criteria, patterns, parameters, indicators and zootechnical characteristics of phenotypic and reproductive value to evaluate queens creates a great void in the exploitation of bees as livestock animals. Consequently, it is necessary to establish practical and easy-to-apply zootechnical methods with the following objectives:

1. Provide the beekeepers with some parameters and patterns that, at the time of the inspection of the hives, allow them to zootechnically describe the queens, in a technical, truthful and fast way; relying on the observation of its most outstanding phenotypical aspects.

2. Make available a technique for quantifying the queen egg laying, based on biological criteria that are easy to understand and apply, with which the beekeepers can evaluate the reproductive behavior of queens within the conception of a farm animal.

References

Shrode, R.R., 1980. Sección IV. Herencia y mejora animal; La herencia y su forma de actuar. En, Curso de zootecnia, editorial Acribia. Zaragoza. España, pp. 253-358.

Sañudo, C., Sánchez, C., Marcén, J.M., 2.009. Capítulo 8. Variación morfológica en bovino lechero. En, Valoración morfológica de los animales domésticos. Realización: SEZ. Coordinador: Carlos Sañudo Astiz. Edita Ministerio de Medio Ambiente y Medio Rural y Marino. Secretaría General Técnica. Centro de Publicaciones, pp. 235-269.

Borror, D.J., Triplehorn, Ch.A., Johnson, N.F., 1989. An introduction to the study of insects. Sext Editions Saunders College Publishing. Harcourt Brace Jovanovich College Publishers. Printed in the United States of America. Library of Congress, pp. 588-664.

Shrimpton, D.H., 1970. Sección II. Especial. La investigación en relación con la ganadería intensiva. En, Zootecnia intensiva, editorial Acribia. Zaragoza. España, pp. 371-613.

Ruttner, F., 2015. Races of bees. In, The hive and the honey bee. Editorial Dadant & Sons. Inc, Hamilton, IL, 62341, USA, pp 47-70.

Daly, H.V., Balling, S.S., 1978. Identification of africanized honey bees in the western hemisphere by discriminat analysis. Journal of the Kansas Entomological Society. 51(4):857-869.

Rinderer, T.E., Sylvester, H.A., Collins, A.M., Pesante, D., 1986. Identification of africanized and european honeybees: Effects of nurse bee genotype and comb size. Bulletin of the Entomological Society of America (32):150-152.

Hoopingarner, R., Farrar, C.L., 1959. Genetic control of size in queen honeybees *J. Econ. Ent.* 52:547-548.

Boch, R., Jamieson, C.A. (1960) Relation of body weight to fecundity in queen honeybees., in The Canadian Entomol. V 92, 9:700-701.

Corbella, E., Gonçalves, L.S. (1982) Relationship between weight at emergence, number of ovarioles and spermathecal volume of Africanized honey bee queens (*Apis mellifera* L.). Rev. Bras. Genet., 4, 835-840.

Costa, F.M. (2005) Estimativas de parâmetros genéticos e fenotípicos para peso e medidas morfométricas em rainhas *Apis mellifera* africanizadas. 39 f. Dissertação (Mestrado em Zootecnia) - Universidade Estadual de Maringá, Maringá.

Halak, A.L., 2012. Parámetros e correlacóes genéticas e fenotípicas para peso e medidas morfométricas em rainhas. *Apis mellifera* Africanizadas. Dissertação apresentada, como parte das exigências para a obtenção do título de Mestre em Zootecnia, no Programa de Pós-Graduação em Zootecnia

Cale, G.H., Sr., Banker, R., Powers, J., 2015. Management of the hive for the production of honey. In, The hive and the honey bee, Editorial Dadant & Sons. Inc, Hamilton, IL, 62341, USA, pp. 463-531.

METHOD TO DETERMINE THE PARAMETER PHENOTYPE OF THE QUEENS

In routine inspections of the hives, beekeepers, whenever possible, look for the queen to make sure of her presence. With this, they check for the same queen and if its appearance is normal.

The objective of this method is to describe the queens based on phenotypic outstanding traits, and as quickly as possible, given their natural tendency of sneaking among the workers during the inspection of the hives and be surrounded and covered by bees. While some queens flaunt themselves, most of them scurry to corners and hide. Moreover, the beekeepers can damage the queens when trying to locate, observe and describe them.

In spite of the fact that many phenotypical measurements can be correlated to queen quality coming from genetic background, developmental, conditions, mating success, beekeeper's management (Hoopingarner & Farrar, 1959; Oldroyd et al, 1990) body weight (Woyke, 1971; Szabo,1973; Szabo et al, 1987 Kahya et al, 2008; Delaney et al, 2011; Tarpy et al, 2011, Collins & Pettis, 2013); when beekeepers are inspecting their hives, they do not have in mind those correlations; they just want to see the queens and describe them. This method claims to be a way to help beekeepers to do so and has no intention to establish any prototype, paratype or morphotype, nor to identify the race to which the evaluated queens belong.

In this method are defined the patterns of the parameter phenotype of the queens, throught their particular physical traits, in order to facilitate their description at the time they are observed during the inspection of the hives. The better the beekeepers have patterns to describe the queens the sooner and preciseer they will be able to do it.

The physical traits of the queens that are used in this method for their phenotypic evaluation have to do with the parameters race (according just to the color), size and age of the queens.

The patterns of these parameters are mainly related to the abdomen of the queens, since due to its dimension is the part of the body where are best appreciated. For the parameter race is considered the color, for the parameter age the color and the physical conditions of the wings and for the parameter size the length of the body from the head to the tip of the abdomen.

The parameter race is assessed through the variable color of the abdomen. There are taken as references the yellow and dark colors, and their variations in greater or lesser degree, of the highest races recognized economic value in commercial beekeeping: *Apis mellifera ligustica* Spin, *Apis mellifera mellifera* L, *Apis mellifera carnica* Pollman, *Apis mellifera caucasica* Gorb and the Africanized honey bee. The hybrid queens coming from the crossing of these races are also evaluated in this method.

When the abdomen is 75% yellow or more, it is considered to be a yellow queen and dark if the color predominant is dark in the same proportion. A hybrid queen will have 50% yellow and 50% dark. Coloration may not be completely delimited because colors may be somewhat interleaved or overlapping but this does not prevent   to define the proportion of colors.

The parameter size is obtained in relation to the variable length of the body, from the head to the tip of the abdomen. It is assumed 20 millimeters for large queens, 15 millimeters for medium queens, and those queens with lower measurement are considered small.

Regarding the age of the queen, it is inferred from the variables degree of brilliance of the abdomen and physical condition of the wings. The abdomen of the workers at birth is pale in color, not so that of the queens that are born with a slightly darker tone, regardless of the race, but in a couple of hours it darkens more and becomes bright. The shine of the abdomen is lost with age, becoming dull and finally discolored. The young queens have a shiny abdomen, the adults have it slightly dull and the old ones discolored. Wings integrity is also used as indicator of age; since in the old queens, they are usually gnawed by wear and rubbing with the worker bees when they move over the honeycombs. However, the nervous nature of some queens can incite the workers to bite the queens' wings, so adult or even young queens can be seen with their wings gnawed prematurely, and this condition can be erroneously attributed to the old age of the queen.

In the parameter phenotype of the queens is have been considered three levels in each pattern to define the possible descriptions that they can have.

Levels

As observing a queen, the first thing that stands out is the color and brightness of the abdomen, then the size and age, likewise the codes that identify the variables color, size and age are arranged. Beside the code that describe the queen Q, the other codes are written: Y= yellow; D= dark; H= hybrid; L=large; M= medium; S= small; Y= young; A=adult; O= old. When looking at a queen there may not be enough time to describe her three body traits, so only one or two variables can be reported. In those cases, the letter I is written instead of the variable(s), which has(have) not been determined. The phenotypic description of the queens gives rise to twenty-nine possible combinations of the color, size and age patterns, nine combinations corresponding to each one (Figure 1).

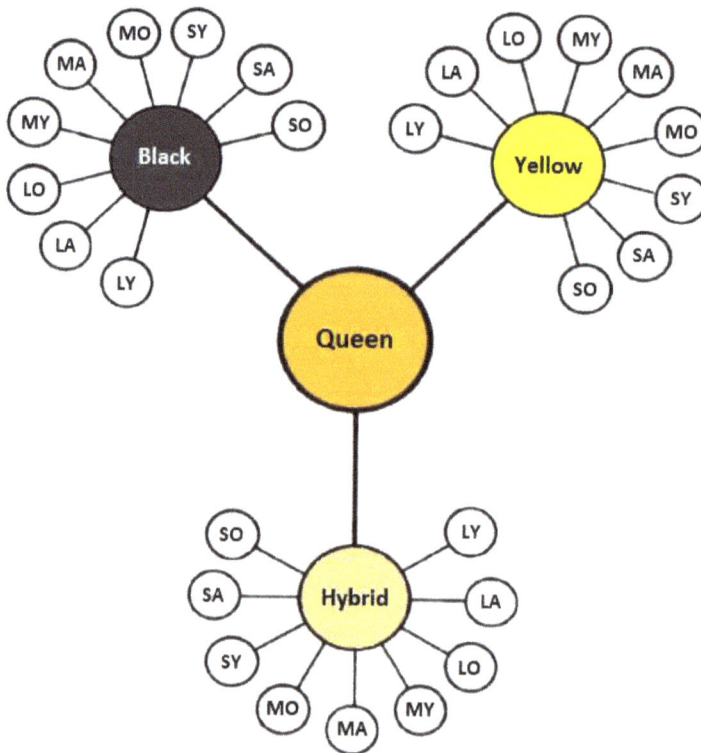

Figure 1. Combinations of the variables of the parameters race, size and age

QYLA	queen yellow, large, adult	QDLY	queen dark, large, young
QYLY	queen yellow, large, young	QDMO	queen dark, medium, old
QHSY	queen hybrid, small, young	QDLA	queen dark, large, adult
QHMA	queen hybrid, medium, adult	QDLO	queen dark, large, old
QDSI	queen dark, small, age indeterminate	QYSO	queen yellow, small, old.
QYII	queen yellow, size and age indeterminate.	QHLA	queen hybrid, large, adult

Figure. 2. Some queens descriptions that can be observed during hive inspections.

Examples of phenotypic descriptions (Figure 3)

Relevance

In the registers of the hives, in addition to the phenotype of the queen, the exact or approximate date of her birth must be noted; or the date of introduction into the hive, if such management was carried out.

The queen is the depository of the genetic material of the colony. She contributes half the chromosome load of the daughters or workers and all the load of the sons or drones. She is, then, the central axis from which all the genotypic, phenotypic and behavioral characteristics of the colony derive.

Hence the importance of knowing exactly the physical aspect of the queen at the moment in which she is observed during a routine inspection of the hive, since its alterations are an unequivocal sign of some biological change that she may have experienced or of some eventuality. These queen phenotypical assessment parameters, on the other hand, help in breeding and selection programs.

References

Collins A.M., Pettis J.S. Correlation of queen size and spermathecal contents and effects of miticide exposure during development. *Apidologie*. 2013; 44:351–356.

Delaney D.A., Keller J.J., Caren J.R., Tarpy D.R. The physical, insemination, and reproductive quality of honey bee queens (*Apis mellifera* L.) *Apidologie*. 2011; 42:1–13.

Hoopingarner R., Farrar C. Genetic control of size in queen honey bees. *J. Econ. Entomol.* 1959; 52:547–548.

Kahya Y., Gençer H.V., Woyke J. Weight at emergence of honey bee (*Apis mellifera caucasica*) queens and its effect on live weights at the pre and post mating periods. *J. Apic. Res.* 2008; 47:118–125.

Oldroyd B.P., Goodman R.D., Allaway M.A. On the relative importance of queens and workers to honey production. *Apidologie.* 1990; 21:153–159.

Szabo T.I. Relationship between weight of honey-bee queens (*Apis mellifera* L.) at emergence and at the cessation of egg laying. *Amer Bee J.* 1973; 13:127–135.

Szabo T.I., Mills P.F., Heikel D.T. Effects of honey bee queen weight and air temperature on the initiation of oviposition. *J. Apic. Res.* 1987; 26:73–78.

Tarpy D.R., Keller J.J., Caren J.R., Delaney D.A. Experimentally induced variation in the physical reproductive potential and mating success in honey bee queens. *Insectes Sociaux.* 2011; 58:569–574.

Woyke J. Correlations between the age at which honey bee brood was grafted, characteristics of the resultant queens, and results

Figure 3. EXAMPLES OF PHENOTYPIC DESCRIPTIONS

1) QYSY (queen yellow, small, yong)

2) QYMY (queen yellow, mediam, young)

3) QYLA (queen yellow, large, adult)

4) QYMO (queen yellow, median, old)

5) QYLO (queen yellow, large, old)

6) QHSY (queen hybrid, small, young)

7) QHSA (queen hybrid, small, adult)

8) QHLA (queen hybrid, large, adult)

9) QHLA (queen hybrid, large, adult)

10) QHLO (queen hybrid, large, old)

11) QDSY (queen, dark, small, young)

12) QDMY (queen dark, median, young)

13) QDMO (queen dark, median, old)

14) QDMO (queen dark, median, old)

15) QDLA (queen dark, large, adult)

METHOD TO EVALUATE THE REPRODUCTIVE PERFORMANCE OF THE QUEENS

This method aims to define some criterias, patterns and a parameter that allow the beekeepers at the time of inspecting a colony, evaluate the queens reproductive performance in a technical, truthful and fast way.

The evaluation of the reproductive performance of the queen refers to the parameter queen egg laying efficiency. This parameter is in function of the bee workers eggs that the queen lays daily and the coverage area that they occupy on the brood chamber combs. The drones eggs are not take into consideration.

It is necessary to assumed the following facts:

a. The Langstroth hive is used.

b. The total area available for the queen egg laying in a brood chamber comb is 4,500 cells on both faces (Fábrega, 1980)

c. In this method is taken as reference a daily egg laying average of 2,000 eggs. It comes from information provided by various authors: Nolan (1925), Laidlaw and Eckert (1962) report that at times egg laying can reach 3,000 eggs / day in European queens; and according to Otis (1991) they rarely exceed 2,000 eggs / day, but Africanized ones can reach 3,000 eggs / day; Harbo (1986) reports 700 and 900 eggs /day in inseminated queens and naturally mated queens and Gary (2015) points to an average 547 eggs / day.

d. The number of brood chamber combs with eggs is estimated according to the Method to estimate the number of frames with eggs, open brood, sealed brood, and bees in the brood chamber at the time of the inspection of a hive (see Chapter two)

To describe the method, it has been elaborated a Table 1, with the criteria, patterns and values that explicate how the parameter is obtained and its interpretation with an daily egg laying average of 2,000, and a second Table 2, with seven daily egg laying averages, to compare them and to give the beekeepers an referential framework to evaluate the queens reproductive performance.

The queen egg laying efficiency is determined by correlating the number of eggs laid in at least three days with the area they occupy in the brood chamber combs. It is assumed four levels of a comb area occupation: abundant egg laying, regular egg laying, few egg laying and null egg laying, and also that the regular egg laying is an half of abundant egg laying, and few egg laying is an half or less of the regular egg laying. Null egg laying means cero.

If there is a comb and a quadrant of another comb or more occupied with eggs, the egg-laying efficiency parameter is named abundant egg-laying, regular egg-laying if there is between one face of a comb and an entire comb occupied with eggs and few egg-laying corresponds to a quadrant of one comb or less occupied with eggs.

Relevance

The queen is the individual around which the functioning of the colony revolves and is supported. She is in charge of producing both, the population of workers and drones; which is directly related to the efficiency of the queens egg laying. A hive, to the extent that it has more worker bees, will have more capacity to produce honey in the harvest season than a hive that has few workers. The latter will also have less chance of surviving in adverse situations or when attacked by predators.

Egg laying is the most important reproductive parameter to assess the quality of a queen. Its purpose is to determine its reproductive efficiency measured by the number of eggs or daily egg laying, reported during routine inspection of the hive. This assessment of the queen will depend only on her, as long as the external factors that condition the egg laying are covered by the beekeeper; such as the presence of sufficient nurses (reinforce the hive when it has a deficit of open and/or sealed brood), space for egg laying and sufficient reserves of pollen and honey. In which case, the reproductive performance of the queen will depend exclusively on her genotype, of the correct anatomical and physiological functionality of its reproductive system and of the quality and quantity of spermatozoa stored in its spermatheca.

The year-round follow-up of the efficiency of a queen's laying offers not only the best assessment of its quality as a breeding caste, but also provides very useful information for queen rearing and breeding programs genetic.

References

Fábrega Roma, A., 1980. Explotación racional del colmenar. **Editorial:** Sintes, Barcelona, España.

Gary, N.E., 2015.Activities and behavior of the honey bee. In, The hive and the honey bee. Editorial Dadant & Sons. Inc, Hamilton, IL, 62341, USA, pp. 247-337.

Harbo, J.R., 1986. Oviposition rate of instrumentally inseminated and naturally mated queen honeybees (Hymenoptera: Apidae). Ann. Entomol. Soc. Am. 79:112-115.

Laidlaw, H., Eckert, J.E., 1962. Queen rearing. Berkeley: University of California. Press.

Nolan, W.J., 1925. The brood rearing cycles of the honeybee. Bull. US. Dep. Agric. N. 1349.

Otis, G.W., 1991. Population biology of the africanized honey bee. In Spivak M, Flectcher DJC, Breed MD, editors. The african honey bee. Boulder C D: Westview, 39:403-423.

Queen egg laying levels					
Cell availability on both faces of a brood chamber comb	Daily egg laying averages	Egg incubation days	Eggs laid in at least three days of egg laying	Combs occupied by eggs in at least three days of egg laying	Parameter queen egg laying efficiency
4,500	2,000	1-2-3	6,000	1.25 a comb and a quadrant of another comb	a ≥ 1.25
	1,000		3,000	0.75 three quarters of a comb	0.50 ≤ r ≤ 1.00
	500		1,500	0.25 one quadrant or less of a comb	f ≤ 0.25
	0				n

The values are approximated to 25%, 50%, 75%, 100%

Table 1. Criteria, patterns and values of the parameter queen egg laying efficiency

Queen egg laying levels with seven daily egg laying averages					
Cell availability on both faces of a brood chamber comb	Daily egg laying averages	Eggs incubation days	Eggs laid in at least three days of egg laying	Combs occupied by eggs in at least three days of egg laying	Parameter queen egg laying efficiency
4,500	2,000 1,000 500 0	1 – 2 - 3	6,000 3,000 1,500	1.25 0.75 0.25	a ≥ 1.25 0.50 ≤ r ≤ 1.00 f ≤ 0.25 n
	2,500 1,250 625 0		7,500 3,750 1,875	1.75 1.00 0.50	a ≥ 1.75 0.75 ≤ r ≤ 1.50 f ≤ 0.50 n
	3,000 1,500 750 0		9,000 4,500 2,250	2.00 1.00 0.50	a ≥ 2.00 0.75 ≤ r ≤ 1.75 f ≤ 0.50 n
	3,500 1,750 0		10,500 5,250 2,625	2.25 1.25 0.50	a ≥ 2.25 0.75 ≤ r ≤ 2.00 f ≤ 0.50 n
	4,000 2,000 1,000 0		12,000 6,000 3,000	2.75 1.50 0.75	a ≥ 2.75 1.00 ≤ r ≤ 2.50 f ≤ 0.75 n
	4,500 2,250 1,125 0		13,500 6,750 3,375	3.00 1.50 0.75	a ≥ 3.00 1.00 ≤ r ≤ 2.75 f ≤ 0.75 n
	5,000 2,500 1,250 0		15,000 7,500 3,750	3.25 1.75 1.00	a ≥ 3.25 1.25 ≤ r ≤ 3.00 f ≤ 1.00 n

The values are approximate to 25% - 50% - 75% - 100%

Table 2. Referential framework to evaluate the queens reproductive performance

Chapter four

MODEL OF ANALYSIS, EVALUATION AND CONTROL OF THE HONEY BEES FOR PRODUCTION PURPOSES

During the routine inspection of the hive, all information of zootechnical character that comes from its biological behavior is gathered and expressed using the Model of analysis, evaluation and control of the honey bees for production purposes, in the form of zootechnical, production, and reproduction parameters, to determine and verify that its development and growth is in accordance with the management, that is being carried out on it. In this way, the beekeepers will be able to correct or repair anomalous situations in the behavior of the colony, that are affecting their production capacity, and animal welfare, in order to obtain the maximum economic benefits (Figure 1).

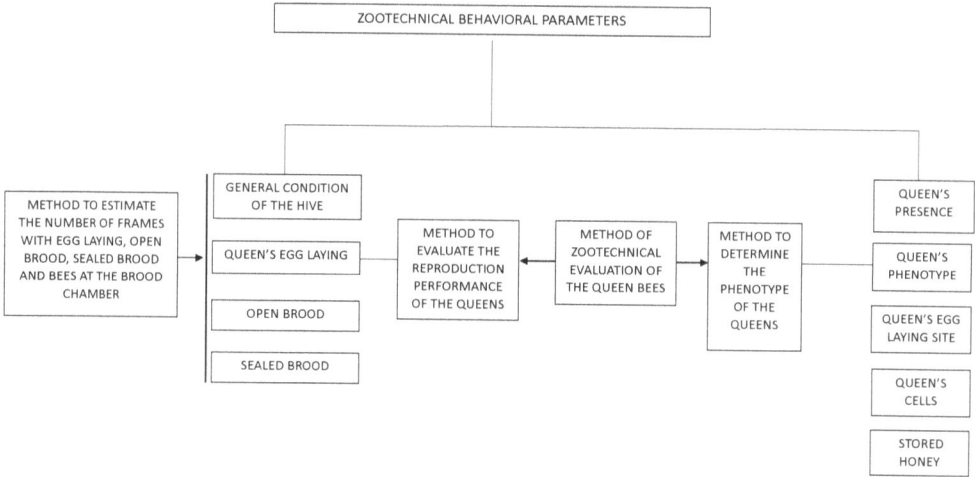

Figure 1. Model of analysis, evaluation and control of the honey bees for production purposes

This model is based on the development and growth of the brood nest and the organization of the population of bees on the honeycombs. Its purpose is to analyze and evaluate the zootechnical behavioral parameters of the colony, not only covering them in entirely, but also in each of its elements separately.

In this model are indicated, fixed and described, using systematic, and methodological criteria, the nine zootechnical behavioral parameters derived from the qualitative and quantitative changes in the egg laying, open brood, sealed brood, and the number of adult bees of a colony, in a Langstroth hive with deep frame type. These parameters have been assigned values that allow them to be compared with the concepts of brood nest and organization of the adult population on the combs, in order to discern their degree of approach to the values of expected productive and reproductive behavior. Likewise, gradation levels have been set, identified with the letters corresponding to the initials of the names of each of them. The parameters that have to do with the phenotype of the queen, its egg laying in the hive bodies, the presence of queen cells and the presence of food reserves in the hive, are also included. Although, the parameters are correlated and   interdependent among themselves, for zootechnical effects their analysis and evaluation is done individually.

THE NINE ZOOTECHNICAL BEHAVIORAL PARAMETERS:

1. General Condition of the hive (Good condition GC, Regular condition RC and Bad condition BC.

2. Queen's presence (Queen's presence QP, Queen's lacking QL and Orphan bee hive OH.

3. Queen's egg laying (abundant a, regular r, few f and null n and has egg laying HE).

4. Queen's phenotype (by color, by side and by age).

5. Queen's egg laying sites (in brood chamber E BC, in brood chamber and super E BCS and only in super E S).

6.   Open brood OB (qualitative and quantitative).

7.   Sealed brood SB (qualitative and quantitative).

8.   Queen cells (open queen cell OQC and sealed queen cell SQC).

9.   Stored honey (mature and immature and quantitative and qualitative).

Utility of the model

1.   Give not only qualitative but quantitative or numerical expression to the biological behavior and activities of the colony, framed as codified parameters; instead of using comparative and superlative terms with which it is not possible to achieve evaluations with values close to the reality of each colony.

2.   Define the parameters which to compare the information gathered from the hives during routine inspections, and discern their degree of approach to the expected production values.

3.   Provide through the interpretation and confrontation of the parameters, a set of solid and verifiable principles that allow to proceed with the bees in a reasoned and logical way.

4.   Dispose of the colony as a single animal entity, but made up of integrated elements, which will allow to evaluated it completely and also individually, and thus decide the most appropriate management to use.

5.   Cover all the organic elements (colony) and hive materials (hive body parts) that are involved in the production process.

6.   Use a technical language, common and accessible to all beekeepers to establish legible and standardized communication channels.

7.   Achieve a strict control of the apiary as a whole and individually.

Considerations to take into account in the model

1. The Langstroth hive is used. This model can also be used in other types of hives, obtaining slight variations that will not alter its applicability on the beekeeping management.

2. The total area available for the queen to lay in a brood chamber comb is 4,500 cells on both faces (Fábrega, 1974)

3. The table Determination of the general condition of the hive (Table 1) is elaborated in function of:

    a) Farrar's rule (number of bees in a hive, their corresponding weight in kilograms, the number of foragers and their percentage according to the bee population (Farrar, 1931):

Number of bees	10,000	20,000	30,000	40,000	50,000	60,000
Number of foragers	2,000	5,000	10,000	20,000	30,000	39,000
Foragers percentage	20 %	25 %	30 %	50 %	60 %	65 %
Bee population weight	1 kg	2 kg	3 kg	4 kg	5 kg	6 kg

    b) The total bee occupancy capacity of a brood chamber comb in the Langstroth hive: 2,430 bees (Burgett and Burikam, 1985)

4. The number of scout bees in the hive at the time of a hive inspection is set at 35% of the total number of foragers. This is obtained from what was pointed out by Winston,1991; Seeley, 1983; Oettingen-Spielberg, 1949 and Lindauer,1952 that scouts are between 5% and 35% depending on the availability of nectar and pollen in the field.

5. For the definition of the parameter queen's egg laying, an average daily of 2,000 eggs is taken as a reference.

6. The determination of the parameters general condition of the hive, queen's egglaying, open brood and sealed brood is carried out fundamentally in the brood chamber. The parameters queen's presence,

queen's phenotype, queen's egg laying sites in the hive, queen cells, stored honey, are obtained from the brood chamber and from the supers according to where they could be found.

Parameter general condition of the hive

It is the first parameter to point out in the routine inspection of the hive. After the number of combs occupied by the bees in the brood chamber has been estimated (by using the Method to estimate the number of frames with eggs, open brood, sealed brood and bees in the brood chamber at the time of the inspection of a hive), the corresponding level of the general condition of the hive is assigned according to the table 1. Determination of the general condition of the hive, that contain the following variables:

1.   Number of bees in the hive

2.   Foraging bees (percentage of 1): 20%, 25%, 30%, 50%, 60%, 65% of the number of bees in the hive.

3.   Scouting foraging bees (percentage of 2): 35% of the number of foraging bees.

4.   Bees in the hive at the time of the inspection: 1 minus 3

5.   Combs in the hive occupied by bees at the time of the inspection: 4 dividing by 2,430

6.   Combs and/or quadrants of the brood chamber occupied by bees at the time of inspection

7.   General condition of the hive.

Levels

Three levels are used to set the values of the general condition of the hive: GC, good condition; RC, regular condition; and BC, bad condition.

Bad condition BC; hives in bad condition are those that have up to four or fewer combs in the brood chamber occupied by bees. Sixteen scenarios are possible.

BC ≤ 4.00

4.00 - 3.75 - 3.50 - 3.25 - 3.00 - 2.75 - 2.50 - 2.25 - 2.00 - 1.75 - 1.50 - 1.25 - 1.00 - 0.75 - 0.50 - 0.25

Regular condition RC; hives in regular condition have in the brood chamber, between four combs and one quadrant of another comb to seven combs and half of another comb occupied by bees. There are fourteen possible scenarios

4.25 ≤ RC ≤ 7.50

4.25 - 4.50 - 4.75 - 5.00 - 5.25 - 5.50 - 5.75 - 6.00 - 6.25 - 6.50- 6.75-7.00 - 7.25 - 7.50

Good condition GC; hives in good condition are those that have in the brood chamber, seven combs and three quarters of another comb or more occupied by bees. Ten scenarios are possible.

GC ≥ 7.75

7.75 - 8.00 - 8.25 - 8.50 - 8.75 - 9.00 - 9.25 - 9.50 - 9.75 - 10.00

Relevance

The population of bees in a colony is in correspondence with the time of year and the efficiency of the queen's egg laying, being the greater or lesser degree of occupation of the bees in the combs, an evident indicative of the degree of strength of the population in a given moment, which also shows the quality of beekeeping management received. According to Farrar, 1931, a strong colony should have 10 brood chamber combs in a Langstroth hive or its equivalent, covered with bees, and colonies should maintain a population of 45,000 to 60,000 bee workers. Furgala, 1975, points out, in turn, that in autumn the colonies with 8 to 10 brood chamber combs can survive the winter and that at the end of this, and early spring the good and productive colonies will be those that have between 15-20 chamber combs (3. 5-5.0 kilos), and that in two months, they can increase to 50,000 or 60,000 bees, and in six more weeks, they can reach 80,000 bees. Thus, in the spring and summer months, the colonies will have the largest number of bees. This is in accordance with the abundance of flowering and consequently with the greatest collection of nectar and pollen to the hive.

The bee population will decrease as flowering begins to decline in autumn, reaching its lowest levels in winter, with around 10,000 bees.

In countries with a tropical climate, on the contrary, the population dynamics is more flexible, in the sense that throughout the year the bee populations remain more vigorous due to the continuous raising of brood; showing, yes, a significant decrease in the periods of greatest scarcity of nectar and pollen (Winston, 1980).

Parameter queen's presence

This parameter deals with the presence or absence of the queen in the hive. Locate the queen can be difficult when there are many bees in the hive, when she is a nervous and elusive, or when she is a newly hatched queen. This parameter should never be missing when the queen is seen, therefore, it will be described covering all its variables; or when the queen has not been seen and there is a clear situation of orphanhood in the hive. The presence of the queen is verified by direct observation of her (whose description corresponds to the parameter Queen's phenotype) or indirectly by seeing the eggs of one day. The absence of the queen is usually evident when the colony is three days orphaned since there are no eggs as no eggs will be found; although there may be times when she is present but without egg laying and there may or may not be queen cells. However, the circumstantial absence of eggs may be a consequence of the momentary shortage of sufficient nurses, or due to the absence of food reserves, it can also happen that after the peak of the development of the brood, the queen decreases the egg laying or restricts it completely. On the other hand, even when the queen is present, that is, she has been seen, there may be very few eggs or not at all, if she is   running out of sperm in its spermatheca, or that, it is a queen with anatomical or physiological malformations. Finally, a queen that has just hatched or is about to start laying after having mated can be seen but there will be no eggs.

Levels

The levels used have to do with the presence or not of the queen. QP, means queen presence and, QL, queen lack, when the queen is not seen and a one day's egg laying neither.

Relevance

This parameter qualitatively expresses the presence of the queen in a hive. It does not include its phenotype neither its egg laying.

The report of this parameter is very valuable when no eggs are seen, since it will have a very different meaning if it is due to the true absence of the queen (orphan hive), or if it does have a queen, but we have not seen it and also there is no egg laying.

Parameter: queen's egg laying and Parameter: queen's phenotype (see chapter three)

Parameter queen's egg laying sites in the hive

This parameter specifies in which body of the hive the queen is laying. If it is only in the brood chamber, in the super (s) or in the brood chamber and in the super(s).

Levels

The levels of this parameter determine the site(s) of the hive where the queen is laying. The letter E, from eggs, means egg laying. The egg laying in the brood chamber is reported with the letters BC, E BC. If egg laying is taking place in the super, it is reported with the letter S, ES. Eggs in the brood chamber and super(s), is indicated as E BC S. The letter S will have beside the number 1,2,3...according the location of each super in the hive.

Relevance

It is very important to locate the hive bodies where the queen is egg laying, as it is indicative of her performance and helps make management decisions:

a) If the queen has an ordered egg laying, she should preferably lays in the brood chamber; so that she will only go up to the super(s) in the following situations:

   1. When the queen does not have combs available in the brood chamber, due to being occupied them by eggs, and/or open and/or sealed brood, or honey and pollen reserves. In such circumstances, these combs will be removed to reinforce colonies that require eggs and/or brood, and they will be replaced with frames (combs) to be used by the queen for lays. Also the combs with reserves could be taken to colonies that need it.

   2. That the queen be disorderly in its egg laying, and consequently lays in the chamber and in the super(s) alternately. Which affects the order and natural arrangement of the brood nest and the way in which bees store honey in the hive.

b) By determining where there is egg laying help to know which body of the hive is capable of providing eggs, open brood and sealed brood in the short, medium and long term, as required, for the orphan hives or for those that need brood and/or young bees.

Parameter open brood and parameter sealed brood

These parameters measure qualitatively and/or quantitatively the worker bees inside the cells after the incubation of the eggs, until the moment they emerge as adult insects.

The amount of open brood and sealed brood are in proportion to the efficiency of the queen's egg laying, as long as they are not affected by circumstances, that may alter the normal development of the life cycle; such as the drastic decrease in the bee population for any reason, or the reserves of pollen and honey in the hive. All of which will result in incomplete development and death of many larvae, also reflected in the decrease in the sealed brood and thus, in the birth of new workers. All of which will negatively affect the survival and production of that colony.

Levels

The levels of the parameters open brood and sealed brood are explained in the chapter five (Beekeeping coding system....)

Relevance

The number of combs with open brood and/or with sealed brood in a hive, allows an estimate of the number of bees that a hive will have in the short and medium term; which will influence its seasonal development. Knowing exactly the amount of open and sealed brood make a good idea of: 1) The degree of potential that the hives will have when the flowering season arrives, and make real projections of the productive performance that can be expected from them. 2) In order to equalize the   hives; the ones that have enough brood to give and those to need   receive. 3) Be clear what hives can be split or take from them   combs to obtain bee nuclei, to increase the bee farm, replace lost hives or sell nuclei. 4) In times of scarcity of nectar and pollen, help to know which hives need to be reinforced and which ones can provide brood, so that the hives can be prepared to cope with the scarcity season that is coming.

Also, to know on the part of the beekeeper of the individual potentiality of each of his hives, in terms of the number of combs with open and/or sealed brood that each of them has, will allow him to know which ones he can split or take from they some combs to obtain bee nuclei, to increase their exploitation, replace lost hives or sell nuclei.

Table 2. Possible occupation degree in the frames of the brood chamber of the variable bees in relation to the variables eggs, open brood and sealed brood.

Parameter queen cells

This parameter has to do with, if there are or not, open, and/or sealed queen cell(s) in the hive. It also indicates the number of them and their location in the combs of the brood chamber and/or in the super(s). The presence of queen cells in a hive indicates that the queen, who has died of any cause, has been or is in the process of being replaced by the colony, and that the colony is requeening, or that the colony is preparing itself to swarm.

Levels

It is written down, if the queen cell(s) is(are) open, OQ, or sealed SQ; their number and if they are in the brood chamber, in the super(s) or in both.

Relevance

The presence of queen cells in a hive is a sure sign of a change in the behavior of the bees towards the queen, as a response to the decrease or total lack of egg laying; either because the spermatozoa in the spermatheca have run out, or because the queen has no space for egg laying, or because the colony is about to swarm or the queen has died.

It is very important to know the number of queen cells, if they are open and or sealed, and their location in each of the hive bodies for the following reasons:

1. By finding the open queen cell(s) and specifying the age of the larva, it can be calculated in how many days the new queen(s) will be born, taking, so, the precaution of reinforcing with sealed brood if necessary and not inspecting that hive on the days when the queen is new born, has reached sexual maturity or is just beginning its egg laying. This is due to the fact that young queens tend to be nervous, which can stimulate their rejection by the workers, and they can be eliminated. In addition, the queen nervousness might cause it to fly out of the hive during the inspection.

2. If there are several queen cells in a colony, some of them can be removed and introduced into another orphan hive(s), which anticipates the birth of the new queen in that hive by several days. This procedure is also very helpful when dealing with orphan colonies that are reluctant to replace their own queen.

3. Knowing the location of the queen cells in the brood chamber combs and/or in a super(s), helps to decide whether to leave them in such comb(s) where were found it, if they were surrounded by sealed and/or open brood, as this guarantees their care and maintenance by the young bees, or relocate them, within the hive, seeking the safety of their preservation until the moment of their birth.

4. Also the location of the queen cell(s) will alert about the care to be taken, when inspecting the body part of the hive, where they are located, so as not to inadvertently damage them during the inspection of the hive.

Parameter stored honey

With this parameter, the presence of mature and/or immature honey in the bodies of the hive will be indicated at least qualitatively, with the exception in harvest times, when it will always be quantitative. In honey harvest seasons, the number of combs found in the super(s) with mature and/or immature honey is recorded. Due to the fact that bees store pollen in less quantity in relation to nectar and that they generally locate it in the same combs where they keep honey, it is considered included within those as food reserves. The honey is stored by the bees in the brood chamber and, as flowering increases, in the supers.

Regardless of the general condition of the hive, this parameter is assumed to be significant to be written down during the hive inspection, the presence of two or more combs with immature or mature honey in the brood chamber, and In the case of the super(s) four or more combs. If there is mature and immature honey in the same super, the one with the most mature honey is noted.

The presence of honey reserve in the brood chamber is registered only in times of scarcity of nectar and pollen in the field.

Levels

See Beekeeping coding system in chapter five.

Relevance

To apply a rational feeding plan to the apiary, the availability of food reserves of each hive throughout the year must be known, in order to specify the program to follow. The hives are naturally fed in times of scarcity of flowering, providing them with sugar syrup, as a source of energy, and soybean meal cakes as a substitute for pollen, which provides the proteins, vitamins, minerals, and lipids necessary for the normal development of the colony.

However, there may be hives with sufficient food reserves to overcome unfavorable environmental conditions, so their feeding would not be necessary at that time. On the contrary, in periods of nectar and pollen flow, some hives in the apiary may have a deficit in reserves, so it will be advisable to feed them artificially.

The control that is had of the alimentary reserves of the hives, will help to decide those that must be fed artificially and in what proportion and those that do not require it; which will result in a better economic management of the resources of the exploitation.

In the honey harvest season it is very useful and practical to have an estimate of the number of frames with mature and immature honey that each super has or the number of supers with ten combs with mature honey in each of the respective hives. Thus, it will be possible to estimate the number of super that will be needed for each hive at the time of harvest; so that the manipulation of the supers with their frames will be streamlined, favoring the work of the beekeeper and benefiting the well-being of the colonies, since the time spent working with them will be reduced, and with it the stress that the entire removal process of honey means for bees.

References

Fábrega Roma, A., 1980. Explotación racional del colmenar. Editorial Sintes. Barcelona, España.

Farrar, Cl., 1931. A measure of some factors affecting the development of the honeybee colony. Unpubl. Thesis. Mass. Sta. Coll.

Burgett, M., Burikam, I., 1985. Number of adult honey bees (Hymenoptera: Apidae) occupying a comb: A standard for estimating colony populations. J. Econ. Entomol, 78:1154-1156.

Winston, M.L., 1991. The collection food. In, The biology of the honey bee. Ed Harvard University Press. Cambridge, Massachusetts, London, England, p. 178.

Seeley, T.D., 1983. Division of labor between scouts and recruits in honeybee foraging. Behav. Ecol. Sociobiol. 12:253-259.

Oettingen-Spielberg, T., 1949. Über das Wesen der Suchbienen. Z. vergl. Physiol. 31:454-489.

Lindauer, M., 1952. Ein Beitrag zur Frage der Arbeitsteilung im Bienenstaat. Z. vergl. Physiol. 34: 299-345. (Transl. Bee World 34:63-73, 85-90).

Furgala, B., 2015 In, Autumn and winter management of productive colonies. In, The hive and the honey bee. Editorial Dadant & Sons. Inc, Hamilton, IL, 62341, USA., pp. 609-632.

Winston, M.L., 1980. Swarming, afterswarming, and reproductive rate of unmanaged honeybee colonies (Apis mellifera). Insectes Sociaux 27:391-398.Table 1. General condition of the hive

GENERAL CONDITION OF THE HIVE						
Number of bees in the hive 1	Foraging bees (percentage of 1) 2	Scout foraging bees (percentage of 2) 35% 3	Bees in the brood chamber at the time of inspection (1 minus 3) 4	Hive combs occupied by bees at time of inspection (4 divided by 2,430) 5	Combs and/or quadrants of the brood chamber occupied by bees 6	General condition of the hive 7
10,000	20% 2,000	700	9,300	4.00	0.25 0.50 0.75 1.00 ... 4.00	4.00 ≤ BC
20,000	25% 5,000	1,750	18,250	7.50	4.25 4.50 4.75 5.00 ... 7.50	4.25 ≤ RC ≤ 7.50
	30% 9,000 50% 20,000				7.75 8.00 8.25 8.50 8.75 9.00 ...	GC ≥ 7.75
30,000	60%	3,150	26,850	11.00	10.00	
40,000	30,000	7,000	33,000	13.50	10.00	
50,000	65%	10,500	39,500	16.25	10.00	
60,000	39,000	13,650	46,350	19.07	10.00	
Note: The values are close to 25% - 50% - 75% - 100%						

Table 1. General condition of the hive

Possible occupation degree in the frames of the brood chamber of the variable bees in relation to the variables eggs, open brood and sealed brood.

Bees %	Eggs %	Open brood %	Sealed brood %
25	25	25	25
	25	25	0
	25	0	25
	25	0	0
	0	25	25
	0	25	0
	0	0	25
	0	0	0
	25	25	50
	25	50	25
	50	25	25
	25	50	0
	25	0	50
	50	25	0
	50	0	25
	50	25	25
	25	75	0
	75	25	0
	0	25	75
	0	75	25
	25	0	75
	75	0	25
	50	50	0
	50	0	50
	0	0	50
	0	50	0
	50	0	0
	0	50	50
	75	0	0
	0	0	75
	0	75	0
	100	0	0
	0	0	100
	0	100	0

Table 2. Possible occupation degree in the frames of the variable bees in relation to the variables eggs, open brood,and sealed brood.

Possible occupation degree in the frames of the brood chamber of the variable bees in relation to the variables eggs, open brood and sealed brood.			
Bees %	Eggs %	Open brood %	Sealed brood %
50	25	25	25
	25	25	0
	25	0	25
	25	0	0
	0	25	25
	0	25	0
	0	0	25
	0	0	0
	25	25	50
	25	50	25
	50	25	25
	25	50	0
	25	0	50
	50	25	0
	50	0	25
	50	25	25
	25	75	0
	75	25	0
	0	25	75
	0	75	25
	25	0	75
	75	0	25
	50	50	0
	50	0	50
	0	0	50
	0	50	0
	50	0	0
	0	50	50
	75	0	0
	0	0	75
	0	75	0
	100	0	0
	0	0	100
	0	100	0

Bees %	Eggs %	Open brood %	Sealed brood %
	25	25	25
	25	25	0
	25	0	25
	25	0	0
	0	25	25
	0	25	0
	0	0	25
	0	0	0
	25	25	50
	25	50	25
	50	25	25
	25	50	0
	25	0	50
	50	25	0
	50	0	25
	50	25	25
75	25	75	0
	75	25	0
	0	25	75
	0	75	25
	25	0	75
	75	0	25
	50	50	0
	50	0	50
	0	0	50
	0	50	0
	50	0	0
	0	50	50
	75	0	0
	0	0	75
	0	75	0
	100	0	0
	0	0	100
	0	100	0

Possible occupation degree in the frames of the brood chamber of the variable bees in relation to the variables eggs, open brood and sealed brood.

Possible occupation degree in the frames of the brood chamber of the variable bees in relation to the variables eggs, open brood and sealed brood.			
Bees %	Eggs %	Open brood %	Sealed brood %
100	25	25	25
	25	25	0
	25	0	25
	25	0	0
	0	25	25
	0	25	0
	0	0	25
	0	0	0
	25	25	50
	25	50	25
	50	25	25
	25	50	0
	25	0	50
	50	25	0
	50	0	25
	50	25	25
	25	75	0
	75	25	0
	0	25	75
	0	75	25
	25	0	75
	75	0	25
	50	50	0
	50	0	50
	0	0	50
	0	50	0
	50	0	0
	0	50	50
	75	0	0
	0	0	75
	0	75	0
	100	0	0
	0	0	100
	0	100	0

Chapter five

BEEKEEPING CODING SYSTEM

Communication and exchange of information between people who performs the same activity requires the use of a standardized, common, easy to understand, logical, written and oral, clear and simple language. Above all, when this activity demands, due to the dynamics of its characteristics, quick and precise actions and responses, as happens with the management of hives in beekeeping operations.

Hence the importance of a technical language based on a coding system, which in addition to improving the fluency and effectiveness of communication between people when working with bees, serves to record in the beekeeping registers the information gathered in the routine inspection of the bees.

The beekeeping coding system consists of a set of codes formed by the acronyms of the beekeeping parameters in their quantitative and/or qualitative expression, the methods and/or techniques used in general beekeeping management, and the situations inherent to the handling of hives.

Beekeeping Coding System

	QUANTITATIVE AND QUALITATIVE BEEKEEPING PARAMETERS - METHODS AND TECHNIQUES - INHERENT SITUATIONS IN THE BEES MANAGEMENT	CODES
1	**TYPES OF HIVES INSPECTION**	
	Complete inspection	C I
	Incomplete inspection	I I
2	**HIVE GENERAL CONDITION**	
	Bad condition	BC
	Regular condition	RC
	Good condition	GC

3	**QUEEN'S PRESENCE**	
	Queen's presence	QP
	Queen's lacking	QL
	Orphan hive	OH
4	**QUEEN'S EGG LAYING**	
	null	n
	few	f
	regular	r
	abundant	a
	Has egglaying	HE
5	**QUEEN'S EGG LAYING SITES**	
	Egglaying in brood chamber	E BC
	Egglaying in super one	E S1
	Egglaying in super two	E S2
	Egglaying in brood chamber and super one	E BC S1
	Egglaying in brood chamber and supers one, two and three	E BC S1 2 3 4
6	**SEALED BROOD (QUANTITATIVE)**	
	There are three frames and a side of a frame with sealed brood in brood chamber	3.50 FSB BC
	Two frames and three quarters of a frame with sealed brood in brood chamber	2.75 FSB BC
	There are four frames and a quadrant of a frame with sealed brood in super one	4.25 FSB S1
	Seven frames and a side of a frame with sealed brood in brood chamber. Three frames with sealed brood in super one	7.50 FSB BC. 3 FSB S1
	Six frames and a side of a frame in brood chamber with sealed brood. Five frames and three quarters of a frame with sealed brood in super one. Four sealed brood frames in super two	6.50 FSB BC. 3.75 FSB S1. 4 FSB S2
7	**SEALED BROOD (QUALITATIVE)**	

	Sealed brood frames in brood chamber	SBF BC
	No sealed brood frames in brood chamber	NSBF BC
	Sealed brood frames in super one	SBF S1
	No sealed brood frames in super two	NSBF S2
	Sealed and open brood frames in brood chamber (there are more sealed brood frames than open brood)	SOBF BC
	No open and sealed brood frames in brood chamber	NOSBF BC
	Open and sealed brood frames in super three	OSBF S3
	No open and sealed brood frames in brood chamber and super one	NOSBF BC S1
8	**OPEN BROOD (QUANTITATIVE)**	
	There are four frames and a side of a frame with open brood in brood chamber	4.50 FOB BC
	Six frames and a quadrant of a frame with open brood in brood chamber	6.25 FOB BC
	There are four frames with open brood in brood chamber and two frames and three quarters of a frame with sealed brood in super one	4 FOB BC. 2.75 FOB S1
	A frame and a side of a frame with open brood in brood chamber	1.50 FOB BC
	There are two frames with open brood in brood chamber, a side of a frame with open brood in super one and a frame and a side of a frame with open brood in super two	2 FOB BC. 0.50 FOB S1. 1.50 FOB S2
	Three frames and a quadrant of a frame with open brood in brood chamber. Four frames and a side of a frame with sealed brood in super one	3.25 FOB BC. 4.50 FOB S1
9	**OPEN BROOD (QUALITATIVE)**	

	Open brood frames in brood chamber	OBF BC
	No open brood frames in brood chamber	NOBF BC
	Open brood frames in super one	OBF S1
	No open brood frames in super two	NOBF S2
	Open brood and sealed brood frames in brood chamber (there are more open brood than sealed brood)	OSBF BC
	No open and sealed brood frames in brood chamber	NOSBF BC
	Sealed and open brood frames in super one (there are more sealed brood than open brood)	SOBF S1
	No open and sealed brood frames in super two	NOSBF S2
10	**QUEEN'S PHENOTYPE**	
	By color	
	Yellow	Y
	Hybrid	H
	Dark	D
	Indeterminate	I
	By side	
	Small	S
	Medium	M
	Large	L
	Indeterminate	I
	By age	
	Young	Y
	Adult	A
	Old	O
	Indeterminate	I
11	**QUEEN CELLS**	

	One open queen cell in brood chamber	1 OQC BC
	Two sealed queen cells in brood chamber	2 SQC BC
	One open queen cell and one sealed queen cell in the same frame in brood chamber. One sealed queen cell in another frame in brood chamber	1 OQC BC1. 1 SQC BC1. 1 SQC BC2
	One open queen cell in super one	1 OQCS1
	Two sealed queen cells in a frame of super three and two sealed queen cells in another frame of super three	2 SQC S3. 2 SQC S3
12	STORED HONEY (QUALITATIVE) MATURE AND IMMATURE	
	There is not stored honey in brood chamber	NSH BC
	Stored honey in super one	SH S1
	No stored honey in super four	NSH S4
	Stored honey in supers one and two. No stored honey in super three	SH S1 2. NSH S3
	There were not inspected the supers	NI S
	There were not inspected super two	NI S2
	Mature honey in supers one, two and three. Immature honey in super four. No honey in super five	MH S1 2 3. IMH S4. NSH S5
13	STORED HONEY (QUANTITATIVE) MATURE AND IMMATURE	
	There are five frames with mature honey in super three.	5 FMH S3
	There are eight frames with mature honey in super one and four frames with immature honey in super two	8 FMH S1. 4F IMH S2
	Ten frames with honey in supers one and two, six in super three and three frames with immature honey in super three.	10 FMH S1 2. 6 FMH S3. 3F IMHS3
	Four frames with mature honey in brood chamber	4 FMH BC
14	CROWDED QUEEN	CQ
15	ATYPICAL EGG LAYING	AE
16	LAYING WORKER	LW
17	QUEEN REMOVED	QR
18	QUEEN INTRODUCED	QI
19	INDIRECT INTRODUCTION OF QUEEN	IIQ
20	BEEKEEPER RELEASED QUEEN	RQ
21	DIRECT INTRODUCTION OF QUEEN	DIQ
22	CROWD OF BEES (LARGE POPULATION OF BEES)	CB
23	LITTLE POPULATION OF BEES	LB
24	ARTIFICIAL FEEDING	

		Syrup fed hive	SFH
		Protein fed hive	PFH
		Hive fed with syrup and proteins	SPFH
25	**TREATMENT**		
		Varroa treatment	VT
		American foulbrood treatment	AFT
		Nosema treatment	NT
26	**BROOD GIVER HIVE**		
		Gave two sealed brood frame of brood chamber to hive 65	G 2 FSB BC H65
		Gave one open brood frame of brood chamber to hive 12	G 1F OB BC H12
		Gave one open and sealed brood frame of brood chamber to hive 35(there are half and half of open and sealed brood)	G 1F OSB BC H35
		Gave one frame and a side of a frame of brood chamber with sealed brood to hive 130	G 1.50F SB BC H 130
		Gave a side of a brood frame of brood chamber with sealed brood to hive 54	G 0.50F SB BC H54
		Gave three quarters of a brood frame of brood chamber with open brood to hive 89	H89 G 0.75F OB BC
		Gave two sealed brood frame of super 1 to hive 7	G 2F SB S1 H7
		Gave a side of a brood frame of brood chamber with eggs and one brood frame and three quarters of another brood chamber frame with sealed brood to hive 1	G 0.50 EF BC. G 1.75F SB BC H1
		Gave one brood chamber frame with eggs and open brood and two brood chamber frames with sealed brood to hive 19	G 1 F EOBF BC. G 2 FSB BC H19
27	**BROOD RECEIVER HIVE**		

	Received one brood frame of brood chamber with three quarters of sealed brood from hive 20	R 0.75F SB BC H20
	Received one brood frame of brood chamber with open brood from hive 40	R 1F OB BC H40
	Received two brood frames and half of another brood frame of brood chamber with open and sealed brood (since open brood predominates over sealed brood, O is written first for open and then S for sealed) from hive 77	R 2.5F OSB BC H77
	Received two brood frames of brood chamber with sealed brood and one with eggs from hive 120 and one with sealed brood from hive 121	R 2F SBF BC H120. R 1F E BC H120. R1F SB BC H121
	Received three brood frames and half of another brood frame of brood chamber with sealed and open brood (since sealed brood predominates over open brood, S is written first for sealed then O for open) from hive 10	R 3.5F SOB BC H10
	Received one brood frame and three-quarters of another brood frame of brood chamber with open and sealed brood (since the open brood predominates over the sealed brood, the O is written first for open and then the S for sealed); and one brood frame and three quarters of another brood frame of brood chamber with open brood from hive 43	R 1.75F OSB BC H43. R 1.75F OBF BC H 43
28	**HONEY GIVER HIVE**	
	Gave five frames with mature honey from super 2 to hive 9	G 5F MH S2 H9
	Gave super 4 with mature honey to hive 55	G S4 MH H55
	Gave five frames with mature honey from super 3 to hive 38	GMHF (5)S3 H38
29	**HONEY RECEIVING HIVE**	
	Received three frames with immature honey in super 1 from hive 24	R 3F IMH S1 H24
	Received super 2 with mature honey from hive 27	R S2MH S2 H27
	Received five frames with mature honey in super 3 from hive 6	R 5F MHF S3 H6
30	**QUEEN CELL GIVER HIVE**	

	Gave one sealed queen cell from brood chamber to hive 17	G 1 SQC BC H17
	Gave one open real cell and two sealed real cells of brood chamber to hive 41(the open real cell is in a frame and the two sealed real cells in another frame	G 1 OQC BC1 H41. G 2 SQC BC2 H41
	Gave one open queen cell and three sealed ones (all three queen cells are in the same frame) to hive 50	G 1 OQC BC H50.G 3 OQC BC H50
31	**QUEEN CELL RECEIVER HIVE**	
	Received one open real cell of brood chamber from hive 14	R 1 OQ C BC H14
	Received one open queen cell and two sealed ones of brood chamber from hive 88 (the open queen cell is in one frame and the two sealed ones in another)	R 1 OQC BC1 H88. R 2 SQC BC2 H88
32	**EXCHANGE OF SUPERS**	
	Super three was exchanged to position one	E S3 1
	Super one was exchanged to position two	E S1 2
	Super four was exchanged to position five	E S4 5
33	**DISEASES**	
	D Foulbrood	D FB
	D Nosema	D N
	D Varroa	D V
34	**QUEEN BORN AND SEEN**	QB S
35	**QUEEN BORN AND NOT SEEN**	QB NS
36	**QUEEN CELLS REMOVED**	
	The beekeeper removed one sealed queen cell in the brood chamber	R 1 SQC BC
	The beekeeper removed one sealed queen cell in a frame and two open queen cells in another frame in the brood chamber	R 1 SQCF1 BC. R 2 OQCF2 BC
	Bees removed two sealed queen cell in a frame and one open queen cell in another frame in the brood chamber	BR 2SQCF1 BC. R 1 OQC F2 BC
37	**SUPERS AND/OR FRAMES OF HONEY HARVESTED PER HIVE**	
	There were harvested supers one and two	H S1 2
	There were harvested supers one and two and seven frames from super three	H S1 2. H 7F S3
	There were harvested five frames from super one, six from super two and eight from super three	H 5F S1. H 6F S2.H 8F S3
38	**HARVESTED FRAMES**	HF

39	**MOVED HIVE**	MH
40	**SPLIT OF HIVE**	
	Split hive	SH
	Nucleus of bees	BN
	inspection of a hive that was split	ISH
41	**INSPECTION OF BEE NUCLEUS OBTAINED IN SPLITTING**	IBN
42	**MISSING HIVE FRAMES**	
	There are missing three hive frames in brood chamber	M 3HF BC
	There are missing two hive frames in super three and one in super one	M 2HF S3.M 1HF S4
43	**SET FRAMES IN HIVE BODIES**	
	It was set one honeycomb in brood chamber	S 1HC BC
	It was set one frame in brood chamber	S 1F BC
	It was set three hive frames in super four and two honeycombs in super one	S 3HC S4. S 2HC S1
44	**REMOVED SUPERS**	
	It was removed super one	R S1
	It was removed super two with frames	R S2 HF
	It was removed super three with honeycombs	R S3 HC
45	**SET SUPERS**	
	It was set super in position two with honeycombs	S S2 HC
	It was set super in position one with frames	S S1 F
	It was set brood chamber with combs	S BC HC
46	**REMOVED OLD OR DAMAGED FRAMES**	
	It was removed one old honeycomb from brood chamber	R 1OHC BC
	It was removed two old honeycombs from super three and one damaged from super four	R 2OHC S3. R 1DHC S4
	It was removed five damaged honey frames from super one	R 5DHF S1
47	**INSPECTION OF ADDED FRAMES**	
	It was not filled a frame of brood chamber	NF 1F BC
	It was filled a frame of brood chamber	F 1F BC
	It was filled three frames of super two and no filled four frame of super three	F 3F S2. NF 4F S3
48	**REMOVED DAMAGED HIVE BODIES**	

	It was removed brood chamber damaged	R BCD
	It was removed bottom board damaged	R BBD
	It was removed inner cover damaged	R ICD
	It was removed super one damaged	R S1D
	It was removed supers one and two damaged	R S1 2D
49	**TRANSFER A NUCLEUS TO A HIVE**	TNH
50	**TRANSFER A HIVE TO A NUCLEUS**	THN
51	**COMBINATION OF HIVES**	
	The hive number five was combined with the hive ten	5C10
	The hive number one hundred was combined with the hive 89	100C89
52	**BROOD CHAMBER WAS NOT INSPECTED**	BCNI
53	**HONEY SUPERS WERE NOT INSPECTED**	
	There were not inspected super one	NI S2
	There were not inspected supers	NI S
54	**DEAD HIVE**	DH
55	**HIVE BODIES**	
	The hive has one brood chamber	H 1BC
	The hive has one brood chamber and one super	H 1BC 1S
	The hive has one brood chamber and two supers	H 1BC 2S
	The hive has one brood chamber and three supers	H 1BC 3S
56	**NEXT INSPECTION**	

Next inspection in three weeks- 15 days- 7days	NI / 3W – NI /15D – NI / 7D – NI / 1M
Watch general condition of the hive	NI / W GC
Watch the queen	NI / W Q
Watch eggs	NI / W E
Watch open brood	NI / W OB
Watch sealed brood	NI/ W SB
Watch queen cells	NI /W QC
Watch honey stored	NI /W HS
Watch crowded queen	NI/ W CQ
Watch atypical egg laying	NI /W AE
Watch laying workers	NI /W LW
Watch queen introduced	NI W QI
Watch artificial feeding	NI W AF
Watch treatment	NI/ W T
Watch varroa	NI/ W V
Watch american foulbrood	NI/ W AF
Watch nosema	NI/ W N
Watch harvested hive	NI /W HH
Watch split hive	NI/ W SH
Harvest hive	NI/ W HH
Split hive	NI SH
Watch nucleus obtained	NI W ON
Watch combined hive	NI W CH
Watch sealed brood frame introduced in brood chamber	NI W FSBI BC
Watch open brood frame introduced in brood chamber	NI W FOBI BC
Watch mature honey frame introduced in super _	NI W FMHI S _
Watch……………………………………………	

METHOD OF HIVES INSPECTION

The method to inspect the hives has to do with the way to gather from the hive that is inspected and record in the field sheet register, all the information that comes from the biological behaviors of the colony, expressed as beekeeping parameters, the different situations and circumstances present in the hive as well as the techniques and/or methods of the general beekeeping performed or to be performed.

Depending on the nine zootechnical behavior parameters, there are two kinds of inspections: Complete Inspection and Incomplete Inspection.

The inspection is complete CI, when the parameters general condition of the hive, queen egg laying, open brood and sealed brood are recorded quantitatively in the field sheet register. Being the incomplete inspection II, if at least one of them is recorded only qualitatively, despite the fact that the other behavioral parameters have been quantitatively noted. However, the general condition of the hive parameter must be noted in the complete and incomplete inspection. It is recommended to carry out at least one complete inspection CI, for every two incomplete ones II, so in this way there will have enough data to make the routine and annual evaluation of the hives.

DEVELOPMENT OF THE METHOD:

In the hive inspection method is described how to search for the information, group it, relate it and transcribe it in the registers, using the Beekeeping coding system. All the information obtained during the inspection of the hives will be fundamental in decision-making in general beekeeping management.

The order of the hives inspection, from the moment the external and internal cover is removed must follow a sequence as a protocol and be executed gradually. The protocol is based on the parameters that are observed according to the arrangement and natural distribution of the eggs laying, open brood, sealed brood and the organization of the bees on the honeycombs.

Once, if the hive has and depending on the time of year, and that the super(s) have been observed to determine the presence of immature and/or mature honey; the brood chamber is inspected looking for: honeycombs with bees, eggs laying, open brood, sealed brood, the queen' egg laying sites, presence of the queen (phenotypically describing it when it becomes visible during the inspection), the possible presence of queen cells and honey reserves.

The order of the information gathered in the field sheet record using the beekeeping coding system is as follows:

Hives inspection type, general condition of the hive, queen's egg laying, queen's presence, queen's phenotype, queen's egg laying sites, open brood, sealed brood, presence of queen cells, presence of honey stored, bodies of the hive, techniques and/or methods used, date of the next revision of the hive and techniques and/or methods to be carried out (if applicable) in the next inspection.

The information collected during the inspection of the hives will be analyzed and evaluated through the Method of analysis, evaluation and control of the honey bees for production purposes, allowing to know the behavior, growth and development of the hives in their zootechnical-productive-reproductive context as organic units, but at the same time differentiating each of its component elements separately.

CONSIDERATIONS WHEN INSPECTING HIVES

1. Starting from the premises regarding the brood nest patterns (see The patterns of the brood nest), the orphanhood of the colony and the laying workers, the hives should be inspected every 24 days. In this way, the beekeeper will find out if the hive has enough brood, young bees, adult bees and reserves of honey and pollen to start requeening queens, if the colony has been orphaned for any reason. On the other hand, the beekeeper will check if there is a population of bees capable of maintaining, protecting and conserving the colony until it is reinforced with honeycombs with eggs or with open brood, which have larvae less than 3 days old-age, to help the colony raise its own queen; or even

better, the beekeeper can proceed to introduce a new queen already fertilized, in order to recover the colony more quickly and bring it back to its stability and functional normality.

2. The frequency of the revisions will be shortened in the cases in which the hives present situations such as those mentioned below:

Orphanhood of the colony.

Orphan colony with queen cells.

Orphan colony without queen cells and which has been reinforced with open brood.

Born queen.

Lack of space for the queen's egg laying for being taken over by the workers.

Few or null egg laying.

Split hive

Pillage or attack by predators on the hive.

Queens introduction.

Harvested hive

Moved hive

RECOMMENDATIONS

The concept of animal welfare must be involved as an unavoidable guideline when working with honey bees. A colony of bees, like any animal, deserves to be treated with the greatest respect and consideration, in the sense of avoiding causing discomfort, disturbance or unnecessary damage; free it from hunger and thirst; prevent and cure their illnesses and provide them with the appropriate hive for their development and survival.

1.   The most convenient type of hive to guarantee adequate animal welfare is the classic Langstroth type vertical hive. It allows the colony to grow in accordance with its natural tendency to lay   in the lower part of the nest, accompanied by some combs with pollen and honey located laterally, the upper part of the nest being where the bees store the bulk of the honey and pollen.

2.   It is recommended that the hives be numbered, starting with the number 1, and so on, trying to maintain the sequence. Enumerate the hives is essential to be able to keep records and to facilitate Zootechnical and General beekeeping management. The beekeepers may have a diagram with the location of the hives of each apiary of the beekeeping operation, which will allow to locate each hive before starting the field work. In this way, beekeepers will not waste time searching the apiary for the hive(s) to be worked on, and   also avoid causing a disturbance in the apiary, by making unnecessary movements among the hives.

     Numbers are advised to be written down in the brood chamber because it is the most visible and durable hive body, taking less damage than flooring, and moving less than the supers or shifts, unless damaged. It is advisable to write the numbers on the front face and on the rear face of the brood chamber, either on the lower right or left, but always following that order, or right or left; so the beekeepers always know where they should look.

3.   The hives must be placed on individual or collective bases about 50 cm high. This seeks to keep the colony away from excessive cold or heat that the soil could transmit to it, thus affecting the development of the young and the incubation of the eggs. Also, high or low soil temperatures will lead the workers to a greater energy expenditure to maintain the temperature of the colony. A separation of approximately 90 cm between hives and 2 meters between rows of hives provides enough space for two operators to work comfortably around a hive, without disturbing or hindering the entry and exit of the foragers from neighboring hives.

     The individual bases have the particularity that they personalize the work with the hive on duty, by avoiding to a certain degree, that the alteration caused by the manipulation of the colony spreads to the neighboring hives, stressing them and arranging them in a defensive attitude to possible intervention back by the beekeepers. Beside, individual bee hive bases reduce the possibilities of pillage that could arise.

In the same way, it should stand out that the use of individual bee hive bases at the recommended height facilitates the handling of the combs and the super(s) and physical favors the beekeepers by having to lean less when working with the hives.

4.   The time spent in inspecting each hive must be as fast and accurate as possible. This is achieved by working with the bees in an orderly, systematic and technical manner, with plans and strategies defined in advance of each visit to the apiary. A correct use of the smoker also influences the greater or lesser degree of damage and stress caused to the adult bees and the brood.

5.   Other details that are included in animal welfare are the proper orientation and location of the hives in the apiary, and the partial shade that the hives must have.

6.   The beekeepers should always bear in mind that every time they inspect a hive are causing a disturbance in the colony, which has a negative impact not only on the adult individuals, but also on the brood. Many bee workers assume attitudes of defense of the nest, with which they stop carrying out their normal activities, affecting the functioning of the colony. Moreover, there is an additional consumption of energy as a consequence of the stress triggered in the population. Open brood and sealed brood can be affected by prevailing environmental conditions and can cause dehydration and chilling, as the case may be, with consequent death.

So, the importance of inspecting the hives as quickly as possible to cause the least disruption to the colony. The inspection time is reduced when the beekeepers have analyzed and evaluated the Individual Hives Register, so that they know in advance which ones require their attention as a priority and which methods or techniques must be applied individually or in groups.

7.   Since the queen is the most important member of the colony, she

requires the most care and attention when hives are being inspected. All manipulation must be done with the certainty of not harming it, for which one proceeds with great caution and delicacy when removing the combs and then replacing them in their place.

8. It is convenient that the inspection of the hives be carried out by two operators, who will be located in a half moon around the hive, without hindering the entrance. The first operator is basically in charge of manipulating the smoker and also to be aware of where his partner, the second operator, is introducing the spatula and helping him with his, to remove the combs. The second operator holds a support table with the Field Sheet Register, where the data of the last inspection made to each hive, that is going to work that day, and the management or technique previously decided if so deserves it. In that same Field Sheet Register, the information obtained from the hive(s) that day and the management or technique that could have been performed is recorded. All the information recorded in the Field Sheet Record will be copied to the respective Individual Hive Register.

OBSERVATIONS TO CONSIDER IN THE HIVES INSPECTION

Among the farm animals, honey bees, due to their condition as insects, are very particular animals. Despite the treatment they have received for thousands of years, they have not become submissive or docile, like the rest of the animals worked by man. They do not recognize the people who usually deal with them, as do the other livestock animals. They have not gotten used to the fact that they will not be harmed by whoever handles them. They are always alert, suspicious and ready to reject the beekeeper's handling. They are animals with a great capacity to perceive through their specialized senses any disturbance in or around their hive, which they interpret as a threat to which they will respond with defensive attitudes, rather than aggressiveness, stinging everything. whatever is close to its hive, without making concessions or distinctions from the beings that are there.

This defensive behavior, more accentuated in some races than in others, is what the beekeepers find every time they open a hive to inspect it and perform any beekeeping technique. But the effects of this manipulation in the hive are more relevant in bees since they entail a progression of events that affect all members of the colony, which will also have an impact on beekeeping production. These events that occur serially are described below:

1. As soon as the beekeepers approach the apiary, the activities of the bees begin to change, especially those in the hives closest to the entrance of the apiary. Restlessness is perceived in the coming and going of the foragers that enter and leave the hives. The guard bees in their entrances begin to release alarm pheromones that spread not only to the interior of each hive but also to neighboring hives. A climate of restlessness is then generated that begins to disrupt the daily work of the bees.

2. In general, guard bees and even some foraging bees that are entering or leaving through the entrances of the hives, initiate the "nest defense" that is to say, to hover and/or sting the beekeeper. These stinging bees die because they are torn apart by the stinger. Death of bees also usually occurs during handling of the hive by crushing with the lid, inter-lid, upper bars of the frames and with the spatula.

3. The alarm pheromones interrupt and disorganize the activities of all the bees, including the queen's egg laying, which can even die from being trapped with the spatula or some frame. The loss of the queen can occur because she flies away from the hive or falls to the ground. The manipulation of the hive involve a generalized stress that harms the colony in its essence as an organic unit that acts and develops as a whole, which leads to stopping or slowing down all the work within the hive, thus affecting the productive process of the colony.

When working with farm animals, to a greater or lesser extent, the degree of stress that can be caused is generally temporary and does not cause significant physical wear and tear on the animals and the productive process inherent to each particular species is little affected.

Another difference when working with bees and doing it with other production animals is that once the work in the apiary begins, it is not possible to take a break as it is possible with other farm animals; since the bees once they begin to manipulate do not give up and continue stinging everything that moves in the surroundings, until even an hour or more after the beekeeper has finished and moved away from the apiary.

Therefore, it is very important that any work with the bees must be carried out as quickly and precisely as possible to trigger the least possible disturbance in the apiary.

BEEKEEPING REGISTERS SYSTEM

The fact of inspecting the hives periodically and collecting the information that comes from the parameters, circumstances of beekeeping nature, techniques and/or methods applied and the work carried out in the beekeeping farms, requires a system of registers. These records must be kept in a systematic, orderly, logical and practical way, in order to give the information, the best applicability for the benefit of the optimal operation of the farm and the beekeeping business.

RELEVANCE OF BEEKEEPING RECORDS

1. The registers allow an individual and collective history of all hives to be kept track of their zootechnical, productive and reproductive behavior. Available at any time for analysis and evaluation, records are essential in beekeeping farms, constituting key pieces for better management.

2. The records constitute a database that facilitates beekeeping work, helping to make the most accurate zootechnical and general management decisions, for the benefit of the success and profitability of beekeeping business.

3. The use of the registers permit to characterize not only each hive individually but also by groups of hives, so that the work in the apiaries becomes easier and faster with its consequent favorable effect on the production and animal welfare of the bees.

    The registers that should be kept in beekeeping farms are of three types:

FIELD SHEET REGISTER

It is the sheet (Figure 1) that the beekeeper takes to the apiary when he is going to inspect hives. The beekeeper has previously read the individual records of hives, analyzed, evaluated and decided on the work individual or collective that is going to perform.

As the beekeeper works in each hive, he writes down all the information that he is gathering from each one of them. At the end of the inspection of the respective hive, he writes down the Code complete inspection (CI) or Incomplete inspection (II) at the beginning of the inspection data for that day, as appropriate. Then all the information will be transferred to the Individual hive register.

FIELD SHEET REGISTER			
INSPECTION N		APIARY NUMBER OR NAME	INSPECTION DATE 03-24- 2022
HIVE	LAST INSPECTION DATE	DATA FROM THE LAST REVISION AND OPERATION CARRIED OUT	NEXT INSPECTION (NI) AND OBSERVATIONS (O)
5	02-28-2022	CI/ GCa/ E BC/ 2.25 FOBF BC/ 3.75 FSB BC/2 F MH BC/6 FMH S1/4 FIH S2/ S S3 HC/ H1BC 3S.	
		New data↓ II/ GC/ E BC/FOBSB BC/ MHS1 2 3/ S S1 F/ H1BC4S	04-10 2022 NI / HH
13	03-15-2022	CI/ RC$_f$ / E BC/ 0.50 FOB BC/ 1 FSB BC/IH BC/ R 1 FSOBF BC H35/.H1BC	03-24-2022 NI/ W GC
		II/ GCr/ E BC/FOB BC/ 2.50 FSBF BC/MH BC / S S1 HC/ H1BC2S	04- 17- 2022 NI/ W E

Figure 1. Field sheet register

INDIVIDUAL HIVE REGISTER

It is the register where everything written in the Field sheet register is transcribed, with their respective dates, management carried out and next inspection. These records are the ones that the beekeeper will analyze and evaluate, and from which he will decide the management plans and/ or general beekeeping techniques to be performed in the apiary in the next routine inspection.

By being clear about the procedure to be performed in each hive, the beekeeper speeds up his intervention and shortens the time to inspect the hive, all of which favors individual and collective animal welfare in the apiary and man-hour efficiency during work.

REGISTER OF ACTIVITIES

It is a booklet where the general activities carried out on the farm are recorded; inspected hives, type of inspection, hives fed and type of food supplied, swarms captured, hive material repaired, number of supers harvested, packaged honey; and any contingency that may occur in the exploitation.

Thus, the use of registers leads to beekeeping operations being managed in an organized, efficient and practical manner in all aspects.

EXAMPLES OF DATA OBTAINED DURING HIVE INSPECTIONS

1) CI / GCa / E BC / 2.25 FOBC / 3.75 FSBC / 2.50 FMH BC / 6 FMH S 1 / IMH S2 3 / S S4 HC/ H1BC 4S.

   Complete inspection. Good condition. Abundant egglaying. Egglaying in brood chamber. Two frames and a quadrant of a frame with open brood in brood chamber. Three frames and three quarters of a frame with sealed brood in brood chamber. Two frames with mature honey in brood chamber. Six frames with mature honey in super one. Immature honey in super two and three (there were not counted the frames). It was set super in position four with honeycombs. The hive has one brood chamber and four supers.

2) II / GC / HE / E BC / OBF BC/ SBF BC/ FIMF BC S1 2/ NSH S3 / G 1 FSOB BC H13 / H 1BC 3S

   Incomplete inspection. Good condition. Have egglaying. Egglaying in brood chamber. There are open brood frames in brood chamber. There are sealed brood frames in brood chamber. Immature honey in brood chamber and supers one and two. There is not stored honey in super three. Gave one sealed and open brood frame of brood chamber to hive thirteen. The hive has one brood chamber and three supers.

3) CI / GCr / E BC S1 / 1.25 FOB BC / 2.50 FSB BC / IMH BC S1 / NSH S2 / H 1BC 2S.

   Complete inspection. Good condition. Regular egglaying. Egglaying in brood chamber and super one. One frame and a quadrant of a frame with open brood in brood chamber. Two frames and a side of a frame with sealed brood in brood chamber. Immature store honey in brood chamber and super one. No store honey in super two. The hive has one brood chamber and two supers.

4) II / GCa / QP / QYLA / E BC / OBF BC/ 4.25 FSB BC/ MH BC S1/ IMH S2 3 / NSH S4 / E S4 1/ H 1BC 4S.

Incomplete inspection. Good condition. Abundant egglaying. Queen's presence. Queen yellow, large, adult. Egglaying in brood chamber. There are open brood frames in brood chamber. There are four frames and a quadrant of a frame with sealed brood in brood chamber. There is mature honey in brood chamber and super one. There is immature honey in supers two and three. Super four was exchanged to position one. The hive has one brood chamber and four supers.

5) CI /BCn/QL/ NOBF BC/ 1.00 FSBF BC/ 1.00 SQC BC/ 1.00 OQC BC/ IMH BC S1/ R1.75 FSB BC H14/ H1BC1S   NI/7D WEL.

Complete inspection. Bad condition. Null egg laying. Queen's lacking. There are not frames with open brood in brood chamber. There is one frame with sealed brood in brood chamber. There is one open queen cell and one sealed queen cell in the same frame in brood chamber. There is mature honey in brood chamber and super one. Received one brood frame and three quarters of another brood frame with sealed brood of brood chamber from hive 14. The hive has one brood chamber and one super. Next inspection in seven days to watch egg laying.

6) CI/ RCn/QL/OH/0.75 FOB BC/3.50 FSB BC/1 OQC BC1/2 SQC BC2/

SH BC S/H1BC1S.

Complete inspection. Regular condition. Null egg laying. Queen's lacking. Orphan hive. Three quadrants of a frame with open brood in the brood chamber. Three frames and half of another one with sealed brood in the brood chamber. One open queen cell in a frame of the brood chamber and two sealed queen cell in another frame of the brood chamber. There are store honey in brood chamber and super. Hive has a brood chamber and a super.

7) CI/GCn/QL/OH/BR 1 SQCF1 BC/ BR 2 SQCF2 BC/4.50 FOB BC/3 FSB BC/5 FMH S1/3.25 FIH S2/SH S3/H1BC3S.

Complete inspection. Good condition. Null egg laying. There is not queen. Orphan hive. The bees removed one sealed queen cell in a frame, and two of another frame of the brood chamber. There are four frames and half of another frame with open brood in the brood chamber. There are three frames with sealed brood in the brood chamber. There are five frames with mature honey in super one and three and a quadrant of another frame frames with inmature honey in super two. There are store honey in super three. Hive has one brood chamber and three supers.

8) CI/GCa/E BC S1/QP/QYLA/3.25 FOB BC/4.25 FSB BC/2.75 FOB S1/3 FSB S1/OBF S2/SBF S2/SH S3 4/S S2 F/G 2F SB BC H88/H1BC5S.

9) II/ RC/ HE/ E BC/ OBF BC/SBF BC/HS BC. S1/ NHS S2/ H1BC 2 S2HS.

10) II/RCa/E BC/2.00 FOB BC/SBF BC/SH BC S1/H1BC1S.

11) II/GC/E BCS1/OBF BC S1/SBF BC S1/SH BC S1 2/ S S3 F/ H1BC 3S NI/3W/WF S3.

12) II/BEa/E BC/4 FOB BC/SBF BC/5 FIM S1/3 FIM S2/ 2 FIM S3/H1BC3S.

13) II/RC/E BC/ 2 FOB BC/3.25 FSB BC/ 6 FMH S1/2.5 FMH S2/ S S3 HC/ H1BC3S.

14) CI/BCn/QL/1.75 FOB BC/2.25 FSB BC/ SH S1 2/NHS S3/H1BC3S/NI 1W/ WE.

15) CI/RCa/E BC/QP/QYLA/1.5 FOB BC/4.25 FSB BC/MH S1/IMH S2/ H1BC25.

16) CI/ GCa/E BC S1 2/4.25 FOB BC/ 3.00 FSB BC/3.75 FOB S1/2

FSB S1/ SH S2 3/H1BC3S.

17)   CI/RCr/E BC/1.50 FOB BC/5 FSB BC/ 4.50 FIMS1/ S S2 F/ H1BC25.

18)   II/BC/HE/ E BC/1 FOB BC/ 2.50 FSB BC/HS S1/H1BC1S

19)   II/GC/HE/E BC/10 FMH S1/ 9FMH S2 3/7 FIMH S4/S S1 HC/H1BC5S
       NI/7D WHH

20)   II/GC/HE/E BC/H F10 S1234/S S1234 HC/H1BC4S

Chapter six

SYSTEM OF BEEKEEPING INDICATORS

To know more precisely the state and development of a bee farm, in addition to having extensive information on Zootechnical beekeeping management and General beekeeping management, expressed through the system of beekeeping parameters and reflected in the registers, it is necessary to resort to beekeeping indicators. The indicators, in addition to serving to analyze and evaluate the effectiveness of beekeeping work, are essential to identify its circumstantial and application failures. The indicators also are required to set the results and goals to be achieved, in order to improve the production and productivity of the farm. All this without forgetting that in an efficient operation there must be an adequate balance between all the elements and factors involved in the production process so that they are never detrimental to the animal welfare of the bees.

Beekeeping indicators can be defined as those characteristic, specific, observable and generally measurable support tools used to detect the changes and progress that occur in zootechnical beekeeping management as a work system and the effectiveness of that management on farms beekeeping.

The beekeeping indicators use the zootechnical parameters to analyze, evaluate and monitor the beekeeping managment status and circumstances; and the parameters of production and reproduction to determine the results of the application of management and thus set the indices or goals to be achieved in the short, medium and long term on the farm. As these are achieved, regional, national and even international indices will be taken as references to improve the benefit and profitability of the farm.

Beekeeping indicators can be qualitative and/or quantitative. The former have to do with the variables presence or absence of parameters or their punctual characteristics and the latter with the quotients that result from the variables of the related parameters. Both types of indicators allow beekeepers to know how the work system is developing and the results that are being obtained in the beekeeping operation, which may or may not be favorable making possible   the pertinent corrections to be made. Indicators must be consistently measurable over time and provide information at relevant intervals.

Zootechnical beekeeping indicators

A) Hives Inspection

1.   Total inspections of all hives (complete and incomplete) TIAH

2.   Complete inspection of all hives CIAH

3.   Incomplete inspection of all hives IIAH

4.   Total inspections of   each hive (complete and incomplete) TIEH

5.   Complete Inspection of each hive CIEH

6.   Incomplete inspection of each hive IIEH

7.   Complete inspection of each hive in relation to its total inspections (complete and incomplete): CIEH X 100/ TIEH

8.   Complete inspection of all hives in relation to their total inspections (complete and incomplete): CIAH X 100/ TIAH

9.   Incomplete inspection of each hive in relation to its total inspections (complete and incomplete): IIEH X 100 / TIEH

10.   Incomplete inspection of all hives in relation to their total inspections (complete and incomplete): IIAH X 100 / TIAH

B) General condition of the hive

1. Hives in good condition HGC

2. Hives in regular condition HRC

3. Hives in bad condition HBC

4. Good condition of each hive in relation to its   total inspections (complete and incomplete): GCEH X 100/ TIEH

5. Good condition of all hives in relation to their total inspections: GCAH X 100/ TIAH

6. Regular condition of each hive in relation to its   total inspections (complete and incomplete): RCEH X 100 / TIEH

7. Regular condition of all hives in relation to their total inspections: RCAH X 100 / TIAH

8. Bad condition of each hive in relation to its   total inspections: BCEH X 100 / TIEH

9. Bad condition of all hives in relation to their total inspections: BCAH X 100 / TIAH

C) Hives with or without queen

1. Hive with queen QH

2. Queenless hive QLH

3. Orphan hive OH

D) Queen's egg laying sites

1. Egg laying in brood chamber in all hives: EBCAH

2. Egg laying in brood chamber in each hive: EBCEH

3. Egg laying in super(s) in each hive: ESEH

4. Egg laying in super(s) in all hives: ESAH

5. Egg laying in brood chamber and in super(s) in each hive: EBCSEH

6. Egg laying in brood chamber and in super (s) in all hives: EBCSAH

7. Egg laying in brood chamber of each hive in relation to its total inspections: EBCEH X 100 / TIEH

8. Egg laying in brood chamber of all hives in relation to their total inspections: EBCAH X 100 / TIAH

9. Egglaying in super(s) in each hive in relation to its total inspections: ESEH X 100 / TIEH

10. Egglaying in super(s) in all hives in relation to their total inspections: ESAH X 100 / TIAH

11. Egg laying in brood chamber and in super(s) in each hive in relation to its total inspections: EBCSEH X 100 / TIEH

12. Egg laying in brood chamber and in super(s) in all hives in relation to their total inspections: EBCSAH X 100 / TIAH

E) Open brood frames

1. Open brood frames of brood chamber in each hive OBFBCEH

2. Open brood frames of brood chamber in all hives OBFBCAH

3. Open and sealed brood frames of brood chamber in each hive OSBFBCEH

4. Open and sealed brood frames of brood chamber in all hives OSBFBCAH

5. Open brood frames of super(s) in each hive OBFSEH

6. Open brood frames of super(s) in all hives OBFSAH

7. Open and sealed brood frames of super(s) in each hive OSBFSEH

8. Open and sealed brood frames of super(s) in all hives OSBFSAH

9. Open brood frames of brood chamber in each hive in relation to its total of open and sealed brood frames: OBFBCEH X 100 / TOSBFBCEH

10. Open brood frames of brood chamber in all hives in relation to their total of open and sealed brood frames: OBFBCAH X 100 / TOSBFBCAH

11. Open brood frames of super(s) in each hive in relation to its total of open and sealed brood frames: OBFSEH X 100 / TOSBFSCEH

12. Open brood frames of super(s) of all hives in relation to their total of open and sealed brood frames: OBFSAH X 100 / TOSBFSAH

F) Sealed brood frames

1. Sealed brood frames of brood chamber in each hive SBFBCEH

2. Sealed brood frames of brood chamber in all hives SBFBCAH

3. Sealed and open brood frames of brood chamber in each hive SOBFBCEH

4. Sealed and open brood frames of brood chamber in all hives SOBFBCAH

5. Sealed brood frames of super(s) in each hive SBFSEH

6. Sealed brood frames of super(s) in all hives SBFSAH

7. Sealed and open brood frames of super(s) in each hive SOBFSEH

8. Sealed and open brood frames of super(s) in all hives SOBFSAH

9. Sealed brood frames of brood chamber in each hive in relation to its total sealed and open brood frames: SBFBCEH X 100 / TSOBFBCEH

10. Sealed brood frames of brood chamber in all hives in relation to their total of sealed and open brood frames: SBFBCAH X 100/ TSOBFBCAH

11. Sealed brood frames of super(s) in each hive in relation to its total of sealed and open brood frames: SBFSEH X 100 / SOBFSEH

12. Sealed brood frames of super(s) of all hives in relation to their total of sealed and open brood frames: SBFSAH X 100 / SOBFSAH

G) Crowded queen QC

    1.   Crowded queen in each hive CQEH

    2.   Crowded queen in all hives CQAH

H) Introduction of queens IQ

I) Indirect queen introduction IIQ

    1.   Indirect introduction of queen in each hive IIQEH

    2.   Indirect introduction of queen in all hives IIQAH

J) Direct queen introduction DIQ

    1.   Direct introduction of queen in each hive DIQEH

    2.   Direct introduction of queen in all hives DIQAH

K) Total of hives combined THC

L) Total of nuclei transferred to hives TNTH

M) Artificially fed hives

    1.   Total of hives artificially fed with syrup

    2.   Total of hives artificially fed with proteins

    3.   Total of hives feeding artificially

N) Number of dead hives

O) Dead hives in relation to the total number of hives in the farm: TDH X 100 / TH

Beekeeping production indicators

    1.   Total of hives TH

    2.   Hives harvested HH

    3.   Honey produced by each hive by harvest: HPEHH

    4.   Honey produced by each hive in all harvests: HPEHAH

    5.   Honey produced by all hives by harvest: HPAHH

    6.   Honey produced by all hives in all harvests: HPAHAH

    7.   Open brood chamber frames given by each hive in relation to its total number of given chamber frames (open and sealed): OBFBCGEH X 100 / (OBFBCG + SBFBCG) EH

    8.   Open brood chamber frames given by all hives in relation to their total number of given chamber frames: OBFBCGAH X 100 / (OBFBCG + SBFBCG) AH

    9.   Sealed brood chamber frames given by each hive in relation to its total number of given chamber frames (sealed and open): SBFBCGEH X 100 / (SBFBCG + OBFBCG) EH

10. Sealed brood chamber frames given by all hives in relation to their total of given chamber frames: SBFBCGAH X 100 / (SBFBCG + OBFBCG) AH

11. Open brood chamber frames received by each hive in relation to its total number of received chamber frames (open and sealed): OBFBCREH X100 / (OBFBCR + SBFBCR) EH

12. Open brood chamber frames received by all hives in relation to their total number of received chamber frames: OBFBCRAH X 100 / (OBFBCR + SBFBCR) AH

13. Sealed brood chamber frames received by each hive in relation to its total number of received chamber frames: SBFBCREH X 100 / (SBFBCR + OBFBCR) EH

14. Sealed brood chamber frames received by all hives in relation to their total number of received chamber frames: SBFBCRAH X 100 / (SBFBCR +OBFBC) AH

15. Open and sealed brood chamber frames given by each hive in relation to its total number of given brood chamber frames: OSBFBCEH X 100 / (OBFBCG+ SBFBCG) EH

16. Open and sealed brood chamber frames given by all hives in relation to their total number of given brood chamber frames: OSBFBCGEAH X 100 / (OBFBCG+ SBFBCG) AH

17. Sealed and open brood chamber frames given by each hive in relation to its total number of given brood chamber frames: SOBFBCGEH X100 / (SBFBCG + OBFBCG) EH

18. Sealed and open brood chamber frames given by all hives in relation to their total of given brood chamber frames: SOBFBCGAH x 100 / (SBFBCG + OBFBCG) AH

19. Open and sealed brood chamber frames received by each hive in relation to its total of received brood chamber frames: OSBFBCREH X 100 / (OBFBCR+ SBFBCR) EH

20. Open and sealed brood chamber frames received by all hives in relation to their total of received brood chamber frames: OSBFBCRAH X 100 / (OBFBCR + SBFBCR) AH

21. Sealed and open brood chamber frames received by each hive in relation to its total of received brood chamber frames: SOBFBCREH X 100 / (SBFBCR + OBFBCR) EH

22. Sealed and open brood chamber frames received by all hives in relation to their total of received brood chamber frames: SOBFBCRAH x 100 / (SBFBCR + OBFBCR) AH

Beekeeping indicators of reproduction

1. Reproductive efficiency of each queen by level (REEQL): it is determined by relating each level of its egg laying LE (a, r, f, n) with its total complete inspection TCIEQ.

    REEQLa = a x 100 / TCIEQ

    REEQLr= r x 100 / TCIEQ

    REEQLf = f x 100 / TCIEQ

    REEQLn = n x 100/ TCIEQ

2. Reproductive rate of each hive; it is determined by relating the nuclei obtained with its total harvests: RREH = NOEH X 100/ harvests

3. Reproductive rate of all hives; it is determined by relating the nuclei obtained with their total harvests: RRAH = NOAH X 100 / harvests

4. Reproductive rate of the Beekeeping farm; it is determined by relating the split hives with the hives in production: RRBF= SH X 100 / HP

Chapter seven

Animal welfare in beekeeping

Animal welfare is a topic that arouses great interest as it is a complex issue with nuances of a social, scientific, ethical, economic, political, cultural and also religious nature.

Regardless of whether we speak of animal law, animal rights or animal welfare separately, together or intertwined, it is implicitly recognized that animals deserve considered treatment *per se* for their capacity to be and feel. Animal rights are current issues, both in the legal field and in civil society, which are increasingly aware of the matter.

Animal welfare has an anthropogenic connotation of a varied and at the same time individualized nature, due to the great animal diversity and the respective *sui generis* particularities of each species. Each person gives a very special and particular value to the type of animal they have, especially those as companions, where dogs and cats stand out, establishing relationships of affection and interdependence that are sometimes very close. These degrees of esteem between man and livestock exploitation animals are not perceived equally or are absent since the tasks and purposes of such animals as well as their behaviors and attitudes do not make it possible. In the case of honey bees there is a treatment of admiration and respect, but always keeping their distance for obvious reasons.

To raise issues regarding animal welfare in beekeeping, the World Organisation for Animal Health (WOAH) must be cited, since it is the intergovernmental entity in charge of ensuring animal health in the world, and to which the World Organization for Trade (WTO) is based on animal health and zoonosis standards. The WOAH publishes two Codes (Terrestrial and Aquatic) that constitute the main references for WTO members. The WOAH's objectives are:

Guarantee the transparency of the animal health situation in the world.

Compile, analyze and disseminate veterinary scientific information.

Advise and stimulate international solidarity for the control of animal diseases.

Guarantee the health security of global trade by developing health rules applicable to international exchanges of animals and products of animal origin.

Improve the legal framework and resources of veterinary services.

Ensure the safety of foods of animal origin and improve animal welfare using scientific bases.

The WOAH carries out its work under the authority and control of a World Assembly of Delegates appointed by the governments of its 182 member countries.

The director general, appointed by the World Assembly of Delegates, directs the activities of the WOAH at its world headquarters, Paris, and implements resolutions drawn up by a Committee supported by five commissions elected by the organization's delegates. These commissions are: the Council, the Regional Commissions and the Specialized Commissions.

From the resolutions issued by the WOAH it is directed the attention to those of the Health Code for Terrestrial Animals (Terrestrial Code) since it contains standards related to animal welfare. The Health Code provides regulatory texts to guarantee safe international trade in terrestrial animals (mammals, reptiles, birds and bees) and their derived products.

The regulations (resolutions) of the Terrestrial Code are found in the following documents: Diagnosis, surveillance and notification of animal diseases; Risk analysis; Quality of veterinary services; Prevention and control of diseases; Trade measures, import and export procedures and veterinary certification; Veterinary Public Health; Diseases common to several species and animal welfare.

Of these regulatory texts, only those related to animal welfare are taken into account for this manuscript. What is specifically related to bee diseases and their prevention and control is leave for specialists in these study disciplines.

WOAH regulations define what animal welfare is and what such a term entails. They, then, describe the basic and scientific principles on which they are based and the criteria for evaluating animal welfare regulations in production systems:

The WOAH designates animal welfare as the physical and mental state of an animal in relation to the conditions in which it lives and dies, which conforms to the so-called "Five Freedoms" globally recognized: living free from hunger, thirst and malnutrition, free from fear and anguish, free from physical and thermal discomfort, free from pain, injury and illness, and free from manifesting natural behavior. Thus, in a general way, the World Organisation for Animal Health postulates that animals must (citing those that have to do with beekeeping operations):

1) Be well fed and in optimal safety conditions, not suffer unpleasant sensations such as pain, fear or restlessness and be able to express behaviors that are important for their state of physical and mental well-being.

2) Have habitability conditions and appropriate sanitary, nutritional and technical management, adequate handling and a safe environment.

3) Define criteria and indicators that help evaluate the extent to which animal management methods influence their well-being. Such evaluations will be individual or in groups using data records.

4) That the comparison of standards and recommendations relating to animal welfare be based more on the equivalence of the criteria to be followed to achieve the objectives of the production systems than on any similarities that may exist.

5) That, by improving the living conditions of animals on farms, production increases and thus economic benefits.

6) Animal handling must promote a positive relationship between humans and animals and not cause injury, panic, lasting fear or avoidable distress.

7) Farm owners and operators must have appropriate knowledge and technical training to ensure that animals receive appropriate treatment.

Specifically regarding animal welfare and production systems, the World Organisation for Animal Health has outlined standards for beef cattle, dairy cattle, working equids, pigs, broiler chickens and killing of reptiles for their skins, meat and other products. For which it points out the common elements in these livestock farms that are taken as foundations to dictate the respective regulations. This is how each exploitation system is defined with its various commercial types and the criteria or measurable variables of animal welfare. And it highlights for all livestock systems the importance of using parameters based on criteria and values with thresholds appropriate to their specific behaviors since they will serve as indicators of good management that must be carried out to guarantee animal welfare on livestock farms.

Unfortunately, the World Organisation for Animal Health has not included honey bees as livestock in its animal welfare standards and guidelines. In this book are developed principles, criteria, patterns, values and variables, parameters and indicators that constitute the platform for working with honey bees within a framework of animal welfare. And also are described considerations and recommendations to follow in routine inspections of hives. All of this help to encourage the animal welfare of honey bees as livestock exploitation animals.

Chapter eight

TRADITIONAL BEEKEEPING

Unlike what happens with other farm animals, honey bees have traditionally been exploited with the criterion that they are animals that do not need much attention and that their use from the productive point of view is achieved with little effort and technical assistance to hives (Whitehead 2014; Bartolini Crespi, 1994; Sepulveda Gil, 1980; Mace, 1974; Harrison et al, 1981).

This pattern of work that prevails in the way most beekeeping operations are carried out is reinforced because honey bees obtain their food from flowers and can reproduce and perpetuate themselves without any human intervention. Hive inspections in traditional beekeeping are notoriously seasonal. The beekeepers inspect the hives three to four times in each season of the year, always depending on the response of the colonies to the environmental conditions and the handling performed in the last inspection, increasing the number of inspections in spring and summer due to the intensity management required by the hives in those months (Cale et al, 2015).

The widespread phenomenon of hive collapse and the problem of the presence of the Asian hornet (Vespa velutina), forced beekeepers to increase the frequency of visits to apiaries to try to control and prevent these problems. In those countries where Africanized bees were established, beekeepers had to implement other techniques and ways of working with bees in order to carry out their beekeeping operations (Michener, 1975; Wiese, 1977; Taylor and Wliamson,1975; Winston,1977; Taylor and Leving, 1978; Rinderer, 1985). However, all these beekeeping management, complementary to the ones used in Europe and the United States of America   actually do not differ in the inappropriate way to evaluate and analyze the hives..

The beekeeper's work has traditionally been to provide his bees with adequate and safe hives to help the growth of the bee population so that they can accumulate sufficient reserves of honey, the main product of the hives, to be harvested. Thus, much of the work with the hives is devoted to placing or removing frames with honeycombs or stamped wax in the brood chamber and/or in the supers, depending on the conditions of the hives and the season of the year (Root, 2013). There are beekeepers who also obtain the by-products of the hives: pollen, propolis, bee nuclei, and even poison; and very few breed queens for their own use or for sale.

During the flowering months, beekeepers usually only inspect the supers to specify the disposition in which the bees are distributing and storing the honey. They usually only go to the brood chamber if the supers have little or no honey, or if they see few bees on the combs or going in and out of the entrance of the hives. They do this to see if there is queen, egg laying, and brood. The rest of the months, in which there is no entry of nectar and pollen or this is not significant, the inspection of the supers is discretionary. The beekeepers inspect the brood chamber to observe egg laying and if possible to see queen; assess the number of combs with open and/or sealed brood and also to remove the damaged combs and clean the bars of the frames.

When the nectar flow is abundant it is very important that hives have enough supers. The first supers can have empty combs or frames with stamped wax which will be carried by the bees to combs that will fill up with nectar (Root, 2013; Cale et al, 2015).

Furgala (2015), points up four fundamental principles of beekeeping management: each colony must have a young queen of proven genetic quality; it must have an adequate reserve of honey and pollen; must be disease free; and must be protected from extreme weather conditions and inhabiting a well-built hive. And so, depending on the beekeeping management carried out with the hives and the observations of the beekeepers during the routine inspections and according to the season of the year, the beekeepers will implement the necessary techniques and/or management methods

It should be noted, however, that in the traditional beekeeping management carried out by beekeepers, there is perceived lack of strategies in the way of collecting the information that comes from the hives during routine inspections, as a consequence, beekeepers reached to wrong conclusions about the conditions in which the hives are, applying inappropriate techniques and beekeeping management. When inspecting hives, beekeepers do not usually follow a protocol of action; they tend to do it messily and randonly. Beekeepers lack of proper ways to collect what they see in hives during inspections, and also lack a frame of reference to help them to analyze and evaluate the information collected.

What do the beekeepers look at first when they start to inspect a hive and how do they interpret it?, what succession of actions are performed so that the information gathered in the inspection of the hives is structured in a logical and orderly manner, so that it can be analyzed with the certainty that it has been well processed in obtaining it, and can lead to a right beekeeping management?

Just as it is necessary for beekeepers to follow a protocol and have a reference procedure when they inspect the hives, they must have homogeneous criteria that allow them to assess everything that happens in the hive in the same way.

Ordinarily, during routine hive inspection, beekeepers check for the presence of the queen either by direct observation of her and/or her egg laying. They inspect the combs in the brood chamber, especially the central ones, to see how many are occupied by egg laying, open and/or sealed brood. To estimate the quality of the queen's egg laying, they use terms such as "very good", "good", "regular "or "bad". However, the same egg laying of a queen can receive a very different assessment depending on which beekeeper it is; thus the egg laying can be very good for one, regular for another, or even excellent for a third beekeeper

Now, what does "very good egg laying " mean ?, what is "a regular egg laying "?, what is "a bad egg laying "? or "an excellent egg laying" ?, on what criteria are beekeepers based to make such assertions? When beekeepers observe egg laying, do they quantify it? If so, how do they do it? under what principles? are beekeepers based on the number of eggs in a comb? how many combs with eggs correspond to a good or regular egg laying, for example? how do they determine all those measurements? what methodology do beekeepers use?

When it comes to describing queens, beekeepers often characterize them using terms such as pretty or ugly, good or bad, young or old, black, very black or yellow, large, medium, or small. Likewise, these definitions of aspects or qualities, ages, colors and sizes of the queens, which can be found as a result of a routine inspection of hives, show a great variety in their conceptualization, according to the criteria of each beekeeper.

Consequently, as beekeepers do not have defined sustainable and commonly used reference frameworks in beekeeping to classify the egg laying, race, age and size of queens, is not possible to achieve rigorous reproductive and phenotypic valorations of them. This is the reason why the evaluations carried out by different beekeepers, when considering the same queens, do not usually coincide.

Other evaluations that beekeepers make when they check their hives have to do with the population of adult bees, the number of combs with open and/or sealed brood, and the store of honey and pollen. Aspects that lead them to determine how a hive is in general at a given moment

Knowing the general conditions of the hives is of the utmost importance, since it shows how the colonies are responding to the prevailing environmental conditions at a certain moment of the annual cycle; especially from flowering changes, if there is a shortage or, on the contrary, enough nectar and pollen in the field. General conditions of the hives are also an indicator of the management that has been applied to them, as a result of decisions taken as a result of previous inspections, which, moreover, can induce the beekeeper to implement in the hives, technical particularities to be performed in the apiary immediately, in the short or medium term.

To assess hives for general condition, beekeepers often classify hives using various terms such as excellent, superior, very good, good, regular, poor, and very poor. This classification is carried out according to the number of bees that observe on the combs at the time of the inspection, the number of combs with egg laying, sealed brood and open brood, and the combs with pollen and honey in the brood chamber and the supers

Regarding the open and/or sealed brood, how does the beekeeper assess this biomass?, based on what criteria does he calculate it?, what method does he use? does he quantify the open and sealed brood, or does he estimate it subjectively? and how does the beekeeper determine if there is a balance between open and sealed brood, and if it is in proportion to the egg laying?, with what do they compare the hives or what reference values do they have to affirm that the colonies improved or worsened or that they remain the same as the last inspection performed on the hives? A beekeeper can even estimate that a hive is in optimal conditions and for others it may be in good or regular condition. From this it follows that beekeepers have disparate estimations of the general condition of the same hive due to the obvious differences in the evaluation criteria.

How does the beekeeper know if the population of bees in a hive has increased or decreased? how does he quantify the number of bees to know if the bee population has increased or decreased? how many combs does the bees occupy?, on the combs can beekeepers estimate increases or decreases in the population of adult bees at a given time? So, there are in the beekeepers notorious differences in appreciation and evaluations of the parameters that are fundamental to evaluate the behavior of the colonies from the zootechnical point of view.

It should also be mentioned that beekeepers do not usually identify their hives with numbers. The purpose of the identification is to know with certainty which hive it is at any time that the beekeepers are in the apiary, or even outside it, when they read the registers to know which hive it is in each circumstance. A hive can not be referred to as "the one with the brood chamber of this or that color, or the one with the stone in the center of the lid is the one that is orphaned or the ones with two small stones are the ones that are going to be harvested". This way of identifying hives, in addition to contravening professional zootechnical management, makes routine work with bees difficult. Likewise, it brings with it confusion when wanting to implement beekeeping registers and is out of keeping with the rigor and seriousness that beekeeping operations must have.

From what has been stated in the previous paragraphs, it can be concluded that the evaluations and analyzes of the hives carried out by beekeepers based on routine inspections of the hives, respond to personal and subjective criteria that are so different and at the same time so individual and specific for each beekeeper; that it is very difficult to find points of agreement that support these evaluation criteria. Consequently, it becomes evident that beekeepers do not have reference frameworks that allow them to compare their observations and estimate the veracity and degree of normality of the biological behavior of the colonies.

All the traditional beekeeping management described in which beekeepers perform routine inspections of hives has, in summary, the following characteristics:

1. Imprecisión and vagueness in the definition of the terms, situations, and development of honey bee hives for production purposes.

2. The routine inspections of the hives and the analysis and evaluation of the information gathered is not based on rigorous zootechnical criteria and biological principles that allow them to be contrasted with the normal functioning of the colonies. These routine inspections lack organization, hierarchy, technique, reasoning and uniformity of criteria, in the way information is collected and evaluated.

3. There is absence of a common beekeeping language, written and oral, clear and simple, that facilitates technical communication among beekeepers, and be used in beekeeping registers. The beekeeping language must be understandable by anyone who works in beekeeping and facilitate a more precise and real view of the general and particular conditions of the hives.

4. Beekeeping registers kept by beekeepers differ in the information they contain and the nomenclature used; since each beekeeper expresses it in his own way, without following codified and permanent drafting rules. Therefore, the information in these registers can only be read and interpreted by   each beekeeper himself.

References:

Whitehead, S.B. 2014. Primera parte. Capítulo primero. Los principios de la apicultura. En, Apicultura moderna, tratado práctico de la cría de las abejas y del cuidado de un colmenar moderno. Editorial Reverte, pp. 5-27.

Bartolini Crespi, A. 1994. Introducción. En, Cría rentable de las abejas. Manual práctico para el apicultor moderno. Editorial De Vecchi, S. A. Balmes, 247.08006. Barcelona, pp. 9-11

Sepulveda Gil, JM.1980. Manejo de las colonias. En, Apicultura. Editorial Aedos. Barcelona. España, pp. 295-311

Mace, H. 1974. Capítulo XXI. Aprovisionamiento de las abejas, pp. 136-139. Capítulo XXIX. Todo el tiempo o parte de él?, pp. 177-180. En, La abeja, la colmena y el apicultor. Segunda edición. Editor José Montesó. Barcelona. Vía Augusta, 251 y 253

Harrison, A.G; Hebden, A; Richard, F.A., FA.1981. Cría de abejas: su miel y sus enfermedades. Editorial Acribia. 1ª edición. Zaragoza. España

Cale, G.H,Sr; Banker, R; Powers, J., 2015. Management of the hive for the production of honey. In, The hive and the honey bee, Editorial Dadant & Sons. Inc, Hamilton, IL, 62341, USA, pp. 463-531.

Michener, C.D. 1975. The brazilian bee problem. Ann. Rev. Ent. 20:399-416.

Wiese, H. 1977.Apiculture with africanized bees in Brazil. Apimondia International Symposium. Pretoria 1976. Am. Bee. J. 117 (3:166-170).

Taylor, O.R; Williamson, G.B. 1975. Current status of the africanized honeybee in Northern South America. Am. Bee. J. 115:92-93,98-99.

Winston, ML. 1977. "The establishment and spread of the africanized honeybees in the Western Hemisphere". J. Ag. Soc. Trinidad and Tobago. 77:306-313.

Taylor, O.R; Levin,M.D. 1978. "Observations on africanized honey bees reported to South and Central American goverment agencies". Bull. Ent. Soc. Amer.

Rinderer, TE. 1985. Africanized honeybees in Venezuela: Honey production and foraging behaviour. Proced. 3 int. Conf. Apic. Trop. Climates. Nairobi. 1984:112-116.

Root AI. 2013.The ABC and X y Z of Bee Culture. Editorial Literacy Licensing. Buenos Aires; pp 244, 245, 246, 247, 381.

Furgala,, B., 2015 In, Autumn and winter management of productive colonies. In, The hive and the honey bee. Editorial Dadant & Sons. Inc, Hamilton, IL, 62341, USA., pp. 609-632.

www.ingramcontent.com/pod-product-compliance
Lightning Source LLC
Chambersburg PA
CBHW050039220326
41599CB00041B/7216